THE UNIVERSE

THE UNIVERSE

Leo Marriott

CHARTWELL
BOOKS, INC.

CHARTWELL BOOKS, INC.
A Division of
BOOK SALES, INC.
114 Northfield Avenue
Edison, New Jersey 08837

ISBN-13: 978-0-7858-2240-0
ISBN-10: 0-7858-2240-2

Printed and bound in China

Design: Danny Gillespie/Compendium Design

Acknowledgments

The History of Astronomy section is taken from articles written by Peter Whitfield for the educational set History of Science and are © Compendium Publishing Ltd. Glossary compiled by Compendium. Unless specified , all images in this book are from NASA's Habble website (http://hubblesite.org). The references below allow the reader to locate the images on the website and gather additional facts, fuller accreditation, and up-to-the-minute information.

1 Hubble 05; 2 2004-20-a; 4 2001-23-a; 5 2000-22-b; 6 NASA; 8 2004-28-b; 9 2004-07-a, 10 AFP/Getty Images; 11 2004-16-c; 12 2004-45-a, 2003-01-a; 14 Compendium, Greenhalf Photography/Corbis; 15 Geeray Sweeney/Corbis, Compendium; 18 Science & Society Picture Library (S&SPL); 19 British Library; 20 S&SPL; 22 Scala/Camposanto, Pisa; 23 Werner Foreman (Field Museum of Natural History, Chicago); 24 Werner Foreman (Private collection, New York); 25 Werner Foreman (Pyongyang Gallery, North Korea); 26 S&SPL; 27 S&SPL; 28 S&SPL; 29 S&SPL; 30 S&SPL; 31 Peter Whitfield collection; 32 S&SPL; 33 S&SPL; 34 British Library, S&SPL; 35 S&SPL; 36 S&SPL; 37 S&SPL; 38 S&SPL; 39 Peter Whitfield collection, 40 S&SPL; 41 S&SPL; 42 S&SPL, Compendium; 43 S&SPL; 44 S&SPL; 45 S&SPL; 46 Compendium, S&SPL; 47 both S&SPL; 48 S&SPL, Compendium; 49 S&SPL; 50 S&SPL; 51 S&SPL, Compendium; 52 Compendium; 53 2002-05-a, Compendium; 54 S&SPL, Compendium; 55 S&SPL, Compendium; 56 Bettmann/Corbis, Compendium; 57 Time-Life Pictures/Getty Images, Getty Images; 58 Time-Life Pictures/Getty Images, Corbis; 59 Bettmann/Corbis; 60 Time-Life Pictures/Getty Images, Getty Images; 61 Compendium, Bettmann/Corbis; 62 Bettmann/Corbis; 63 Bettmann/Corbis; 64 Compendium, 2004-23-a; 65 Compendium; 66 S&SPL; 68 Compendium; 69 S&SPL, Compendium; 70 Compendium; 71 Compendium; 72 Compendium; 74 Hubble 06; 76 Hubble 05; 78 Hubble 11; 79 Hubble 17; 80 Hubble 15; 81 Getty Images; 82 Getty Images; 83 Getty Images; 84 NASA; 85 Bettmann/Corbis; 86 NASA/Roger Ressmeyer/Corbis; 87 NASA/Roger Ressmeyer/Corbis; 88 NASA/Roger Ressmeyer/Corbis; 89 both NASA/Roger Ressmeyer/Corbis; 90 NASA; 92 1991-09-b; 93 1991-13-a; 94 1991-05-a; 95 1994-36-a; 96 2001-15-c; 98 2005-17-b; 99 1999-14-c; 100 1995-18-b, 1994-35-a; 101 2004-30-a; 102 1992-24-a, 2003-17-b, 2001-24-a, 2005-16-b; 103 1998-18-a, 2005-06-d, 1990-11-a; 104 2003-28-a; 106 2005-11-a; 107 2005-12-a; 108 2005-09-a; 109 2005-04-a; 110/1 2004-15-a, 2004-15-b; 112 2003-24-a; 113 2002-11-a; 114 2004-31-b; 115 2004-04-a; 116 2001-37-a; 117 2004-13-e; 118 2005-01-a; 119 2005-01-a; 120 1994-25-a, 2000-34-a; 122 2001-10-a; 123 2002-21-a; 124 1999-44-a; 126 2003-14-a; 127 1992-17-a; 128 1998-14-c; 129 2002-22-a; 130 2001-22-a; 131 2003-15-a; 132 1992-27-a; 133 1999-19-b; 134 2001-28-b; 135 1992-01-a; 136 1999-04-e; 137 1997-34-d; 138 2001-32-b; 139 1995-07-a; 140 2003-04-a; 141 1993-23-b; 142 2005-05-a; 143 2003-18-a; 144 2003-01-a; 145 1994-40-a; 146 1999-25-a, 1996-11-a; 148 2000-14-a; 149 2000-37-a; 150 2001-01-a, 2001-02-a; 152 2001-26-a; 153 2001-16-a; 154 2001-23-a; 156 2004-25-a; 157 2002-29-a; 158 2002-07-a; 159 2002-03-a; 160 2003-11-a; 162 2003-11-b; 164 2004-32-d; 165 2004-27-a; 166 2005-12-b; 168 2002-11-f; 169 2003-16-a; 170 2002-11-c; 171 2002-11-g; 172 1995-45-a; 173 2004-17-c; 174 1999-35-d; 175 1994-10-a; 176 2005-02-b; 177 2003-20-a; 178 1999-42-a; 179 2002-01-a; 180 2000-07-a; 181 2001-21-a; 182 2000-04-a; 184 1999-35-c; 185 2002-14-a; 186 2002-15-a, 2002-19-a; 188 2000-06-a; 189 2004-22-c; 190 2001-12-a; 191 2001-13-a; 192 2002-25-a; 193 1990-26-a; 194 1992-29-b; 195 1992-12-a; 196 1992-08-a; 197 2005-15-f; 198 2002-24-a; 199 1993-02-a; 200 1993-01-a; 201 1996-13-a; 202 1990-09-e; 203 2003-10-I; 204 1999-33-d; 205 1995-11-a; 206 1995-44-a; 207 2004-10-f; 208 2000-23-b; 209 1996-38-a; 210 2003-31-b; 211 2003-06-d; 212 2003-31-b; 213 1993-23-b; 214 1994-24-a; 215 1999-20-a; 216 2002-05-a, 2000-10-a; 218 2004-26-a; 219 2000-28-a; 220 2001-11-a; 221 2000-30-a; 222 2001-34-a; 223 2001-39-a; 224 1993-11-b; 226 2004-20-a; 228 2004-29-I; 229 2004-35-a; 230 1991-03-a; 231 1991-12-a; 232 1993-10-a; 234 1999-08-a; 235 1990-15-a; 236 2001-21-a; 238 1999-20-a; 239 2001-25-a; 240 1993-13-a; 241 2003-19-I; 242 2000-33-b; 243 1999-26-a; 244 2005-10-b; 246 1994-02-b; 247 2004-34-c; 248 2005-15-a; 249 2004-06-a; 256 2005-15-c.

PAGE 1: The Hubble Space Telescope.

PAGES 2/3: The Hubble Space Telescope's Fine Guidance Sensors have refined the distance to the Pleiades at about 440 light-years. This color-composite image of the Pleiades star cluster (see also pages 226–7) was taken by the Palomar 48-inch Schmidt telescope. The image is from the second Palomar Observatory Sky Survey. It was made from three separate images taken in red, green, and blue filters.

LEFT: An unusual edge-on view of the unusual twisted disk structure of galaxy ESO 510-G13. The warped dusty disk is visual evidence of how colliding galaxies spawn the formation of new generations of stars.

RIGHT: Stephan's Quintet is a collection of very active galaxies at least two of which have ripped stars and gas from neighboring galaxies and tossed them into space. But this violence has spawned over 100 star clusters—each containing millions of stars—plus several dwarf galaxies.

PAGES 6/7: Sunrise over Earth—a NASA photograph.

Contents

Introduction

Since time immemorial mankind has gazed upwards at the sky and heavens above, trying to make sense of the objects and phenomena observed. From our viewpoint as supposed intelligent, knowledgeable and civilized people, we are quick to seek and ascribe a logical scientific explanation to any new or unusual occurrence. However our early ancestors, whose lives centered solely around a perpetual battle with the forces of nature to maintain their precarious existence, were more profoundly affected. They were quick to associate astronomical happenings with incidents and setbacks in their own circumstances. The basic human fear of the unknown led to phenomena such as comets, meteorite showers, and eclipses often being seen as harbingers of doom. The sun and the moon, as the most significant objects in the sky were endowed with all kinds of attributes and in many instances were revered as gods and deities.

However another basic human instinct—curiosity—enabled people to begin to make some sense of what they could see. Certainly the rise and set of the sun provided an indication of daily time; the length of the day was related to the seasons, as was the changing pattern of the stars. It was also realized that while most stars were in a fixed position relative to the others, a few seemed to trace their own path through the night skies. These so-called wandering stars were, of course, the planets, the name coming from the Greek word *planetes*—literally wanderers. The waxing and waning of the moon was seen to influence the rise and fall of tidal waters.

Although the Greeks were among the earliest to set down the results of their observations and to produce the first cogent theories to explain what they saw, there is no doubt that earlier peoples already had the facility to make some quite complex astronomical observations. Certainly many Neolithic structures, of which Stonehenge is a prime but not unique example, could only have been erected with the benefit of an ability to forecast the movement of celestial bodies. Nevertheless it was the work of Greeks such as Eudoxus, Eratosthenes, and Hipparchus who laid the foundations of modern astronomy and began to understand the workings of the universe. They created the concept of heavenly bodies moving in circular or elliptical paths, measured the circumference of the Earth (accepting in so doing that the Earth was a sphere), and made countless observations which were diligently and accurately recorded. The observations were later used by Claudius Ptolemy to produce *The Almagest*, widely regarded as the most important astronomical work of antiquity. Admittedly in Ptolemy's universe the sun, moon, and planets revolved around the Earth which involved some abstruse mathematics to make the known facts fit the theory. Nevertheless it was the first significant attempt to

LEFT: A deep space view known as the Hubble Ultra Deep Field (HUDF), looking at objects between 400 and 800 million light-years away. This was taken using the Advanced Camera for Surveys (ACS) and the Near Infrared Camera and Multi Object Spectrometer (NICMOS), installed during the 2002 service mission. The ACS has twice the field of view of the Wide Field Planetary Camera 2 installed in 1993 and which has itself produced some fascinating images and is considerably more sensitive. The deep space image shown here picks out an area of space in the Fornax constellation which to Earth-based observers appear relatively clear. The whole image was built up from a series of observations from September 2003 to January 2004 and the photos of light captured from these distant objects started their journey through space even before Earth existed—a sobering thought!

FAR LEFT: Detailed analyses of mankind's deepest optical view of the universe, the Hubble Ultra Deep Field (HUDF), by several expert teams have at last identified what may turn out to be the earliest star-forming galaxies. Astronomers are now debating whether the hottest stars in these early galaxies may have provided enough radiation to "lift a curtain" of cold, primordial hydrogen that cooled after the big bang. This is a problem that has perplexed astronomers over the past decade, and NASA's Hubble Space Telescope has at last glimpsed what could be the "end of the opening act" of galaxy formation. These faint sources illustrate how astronomers can begin to explore when the first galaxies formed and what their properties might be. But even though Hubble has looked 95 percent of the way back to the beginning of time, astronomers agree that's not far enough.

provide an explanation for the movements of the solar system and was accepted until well into the sixteenth century.

In fact it was not until that time when European explorers began to open up the world and voyages of circumnavigation were made that it was finally accepted that the Earth was in fact a sphere, despite the earlier reasoning of the Greek philosophers. The exploration of the seas required accurate navigation and stellar observations were to be of the utmost importance until the development of radio-based navigation systems in the twentieth century. It was in the sixteenth century that Nicolas Copernicus postulated the so-called heliocentric hypothesis in his book *De Revolutionibus*. The concept that the Earth and other planets revolved around the sun is taken for granted today but at the time it was revolutionary, and even heretical, and was not immediately accepted. However his theory was backed up by observations made by Galileo and in the seventeenth century Johannes Kepler was able to formulate the basic rules of planetary motion. He was followed by Isaac Newton who gave substance to these rules, and explained the relationship between all bodies in the universe by combining the outcome of mass, momentum and force into the concept of gravity.

The great boost to astronomic theory and observation was the invention of a practical telescope by Galileo, enabling objects not previously visible to the unaided naked eye to be seen and cataloged for the first time. This led to the discovery of the moons of Jupiter, the planet Uranus and, following calculations based on Newton's work, the planet Neptune. In the nineteenth century, as the optical quality of telescopes was constantly improved, the invention of associated instruments such as spectrometers and cameras opened up whole new fields of study and enabled significant information about the nature of planets stars and galaxies to be gathered. Optical technology advanced in leaps and bounds in the twentieth century resulting in installations such as the 100-inch Hooker telescope at Mount Wilson and the 200-inch Hale telescope which enabled galaxies and other phenomena to be examined in much greater detail. In the 1930s a new science of radio astronomy was born and this led to the detection of new objects such as pulsars and quasars, although such discoveries raised more questions than they answered.

However the greatest advances in astronomic sciences has come about with mankind's successful forays into space coupled with the staggering advances made with digital computer technology. The combination of these developments culminated in the launch of the Hubble Space Telescope (HST) in 1992. Previously all optical observations had suffered from the fact that the light reaching the telescopes had to pass through the Earth's atmosphere, so that the final image was subject to various distortions as well as being fainter than would otherwise be the case. With the coming of the HST coupled with digital camera technology, observers were able to see undistorted images for the first time and to reach far deeper into space than had ever been the case in the past. Almost all of the images in this

book have come from this source and the reader can only marvel at the scope and diversity of the results.

In trying to observe and understand the universe, the average person is quickly overwhelmed by its sheer size and scale. We glibly refer to objects being located so many light-years away from Earth and this gives us a reassuring unit of measurement. However, when it is realized that light travels at 186,000 miles per second , the distance covered in only one light-year is unimaginable. Even in one day, light travels over 16 billion miles. When we talk about some objects being 15,000 billion light-years away (the furthest observed by Hubble), the final figure could probably not be written out on this page!

This book is divided into opening historical sections that look at the history of astronomy and at Hubble himself, before moving on to show the remarkable visions of the galaxy that the HST has given us.

ABOVE: The US space shuttle *Columbia* clears the tower early on March 1, 2002 at launch pad 39-A of the Kennedy Space Center in Florida. The *Columbia* and its crew were beginning an eleven day mission to service the Hubble Space Telescope. AFP PHOTO/Bruce WEAVER

RIGHT: In an ambitious project to bring the capabilities of the HST to a wider audience, a team of experts at the Space Telescope Institute created a three-minute film simulating a voyage through the universe and its galaxies. More than 11,000 galaxy images were extracted from those obtained in the Great Observations Origins Deep Survey (GOODS) and manipulated into an incredible 3D model. Projected on a massive IMAX screen, it does more than just journey through space. Due to the distance of the galaxies shown, it is a voyage back in time to the early origins of the universe.

LEFT: Detail of the Abell 1689 galaxy cluster. The cluster is so vast that it distorts the images of the galaxies behind it through the influence of gravity and dark matter. At two-million light-years wide this "gravitational lens" is a phenomenon that creates multiple images of the individual galaxies.

RIGHT: A wonderful view of two galaxies (NGC 2207 and IC 2163) in the process of colliding. The whole process will take millions of years but scientists speculate on what happens in such situations as previous such occurrences have shaped many of the galaxies which we can see today. A subject of particular speculation is what happens when such colliding galaxies contain black holes at their center. It is thought possible that the immense gravitational effects could dislodge the black holes so that in extreme conditions they could be thrown completely clear although more likely is that they would be displaced from the center, and then gradually drift back again to their original position at the center of the galaxy. Astronomers are therefore keen to locate a galaxy with an offcenter black hole which would confirm this theory, although none have been found to date.

The History of Astronomy

STONES AND STARS: ANCIENT MEGALITHIC ASTRONOMY

The Old Stone Age gave way to the New approximately 10,000 years ago, when the peoples of Europe and Asia experienced a major enrichment of their cultural life. This enrichment followed the emergence of agriculture—the raising of crops and the breeding of animals, which together gave man a controlled food supply. That permitted them to abandon life as wandering hunters and to create settled communities, first in villages and then in the earliest cities.

Agriculture seems to have emerged in southwest Asia and to have spread steadily across Europe until, by 4000 B.C., it had reached France, Britain, Spain, and southern Scandinavia. Agriculture and the rise of villages and cities produced enormous consequences for social life: division of labor and new patterns of work, new political structures, and new technologies. These technologies included pottery, textiles, metalworking, and eventually the discovery of the wheel.

Food production changed with the discovery of breadmaking, the properties of salt, and the ability to store food. The horse was tamed, and wheeled carts were constructed.

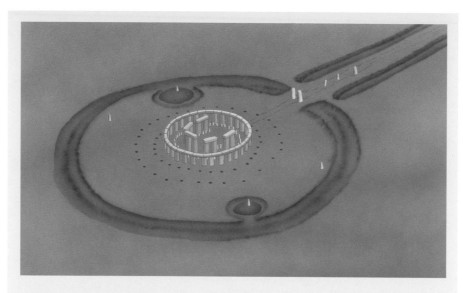

STONEHENGE

The most famous of all the megalithic sites is Stonehenge in Wiltshire, England. Now believed to have gone through several phases of development between 3000 and 2000 B.C., the whole site is aligned from the northeast to the southwest, toward the rising Sun on midsummer day and the setting Sun on midwinter day. Outside the main circle further stones appear to mark the highest and lowest points of the Moon's rising and setting. Other alignments to stars have also been suggested. It was not, as has often been claimed, an ancient observatory for studying the heavens, but like Newgrange, it did embody such observations. Once again, the design of the stone circles seems to have been intended to direct and trap light between stones for brief moments of the year. But what actually happened at Stonehenge? What rituals were enacted there? Why did the people of ancient Britain labor for years to erect this massive structure? Why was the site abandoned around 1000 B.C., so that when the Romans arrived in Britain, it was deserted and its purpose long forgotten?

We will never know the answers to these questions, but what is certain is that for centuries the people of Stonehenge watched the stars with scientific precision. They mastered the movements of the heavenly bodies, and they incorporated that knowledge into a system of beliefs about the universe. They built up a philosophy of nature based on observation of nature's workings. They left no written records for us to study, but they left probably the oldest records of precise scientific activity. That is why the history of recorded science begins with the stone circles of Neolithic Europe.

BELOW LEFT: Trethevy Quoit, Cornwall, England. Also called dolmens or cromlechs, quoits were probably covered by an earth mound providing a closed chamber for a burial.

LEFT: A reconstruction of Stonehenge in its final phase, about 1550 B.C.

BELOW RIGHT: Burial chamber at West Kennet, Wiltshire, England. The lines of sight are limited by the edges and faces of stones and thus allow alignment with astronomical bodies at specific times of year.

FAR RIGHT: Newgrange is one of the largest Neolithic passage tombs in Europe.

PAGE 16

ABOVE: A copy of the hieroglyphics and other figures embossed on the Karnak "Clepsydra" (water clock), providing an Egyptian celestial diagram dating back to around 1400 B.C. The top row shows a series of planet gods and the thirty-six decan stars, the great celestial timekeepers of the ancient Egyptians. In the middle are various constellations and deities. The bottom row shows the calendar of months and the monthly gods.

BELOW: Egyptian water clock dating to 1415–1380 B.C. This is a plaster cast of an original from Karnak temple. In use the alabaster original was filled with water, which leaked out of a small hole in the

PAGE 17

ABOVE LEFT: The Babylonian Sun Tablet of Sippar. The huge figure on the right is the sun-god Shamash. The king and two priests watch as an image of the Sun is placed on the altar, ninth century B.C.

BELOW RIGHT: The Aristotelian cosmos of seven planets, including the Sun and Moon, a ring of stars, and Earth in the center.

Secrets in stone

Just as people's material life was becoming richer and more complex, so we have clear evidence that their intellectual life was developing too. In particular it seems that their understanding of nature was becoming more precise, and that they were building up belief systems about the universe. Neolithic people were still preliterate, so we have no written statements about their knowledge or beliefs, but they nevertheless gave them lasting form in the stone monoliths that they built.

Northwestern Europe is full of stone avenues and stone circles that have been built in such a way as to show that the people who put them there between three and six thousand years ago were expert astronomers who had mastered the periodic movements of the Sun, Moon, and stars by watching them intently over many centuries.

Some of these megaliths are as old as the pyramids of Egypt and the ziggurats of Babylonia, and they show that the peoples of Neolithic Europe shared some of the intellectual qualities of the Near Eastern civilizations. Like the pyramids, the stone circles are important because of their huge scale: As feats of engineering created by the strength of human hands, they represent a colossal investment of time and effort. The large sites such as Stonehenge must have demanded millions of hours of labor. Surely only the most powerful of motives can have inspired such projects, so that they must provide us with valuable insights into the mind of Neolithic peoples.

Irish Tombs

At Newgrange in Ireland a row of tombs was built from massive stone slabs, all facing a long, curved passageway. For just a few days around the midwinter solstice the light of the rising Sun enters the passageway through a small opening at the entrance and lights up the tombs. The entire tomb structure was designed and built to capture this astronomical moment, but why we do not know.

Were the people buried there special? What was the association between the spirits of the dead and the sunlight at its lowest rising point of the year? Was some force passing into the tombs, or perhaps out of them? It used to be thought that such structures must be calendars of some kind. But why build such an elaborate structure when a few wooden posts would have worked just as well? And why bury the dead in such a place? No, structures such

as Newgrange were not astronomical calendars, but they did embody calendrical and astronomical events that had already been worked out, and that were then built into the structure. It seems probable that this form of astronomical monument was part of a system of religious beliefs in which the heavenly bodies were somehow directly linked to the existence of human beings on Earth, although we cannot say how.

Aligned to the Stars and Moon

This religion—if that is what it was—was not confined to the Sun. In a Neolithic burial chamber of West Kennet in Wiltshire, England, there are five tombs containing the remains of forty-six people. Each of these tombs has been aligned to receive light from the rising of a bright star such as Betelgeuse, Spica, or the

Pleiades in the positions in which they stood around 3500 B.C.

In northeast Scotland there is a distinctive group of stone monuments built in the pattern shown above right. They face other upright stones nearby. Experts have discovered that these stones marked the highest or lowest points in the Moon's rising

and setting, points that recur in a cycle of eighteen years. The observations that lie behind these stones must have occupied many decades, even centuries, but no physical traces of such observations have ever been found.

ANCIENT ASTRONOMY
Calendars

The oldest exact science, in which laws and rules could be deduced from careful observation, was astronomy, and Egyptian astronomy combined exact science with certain mythical beliefs. The first practical motive for astronomy was calendar making. The cardinal points of the year for northern peoples—the Sun's solstices and equinoxes—were of no special importance to the Egyptians at the latitude of 25 to 30 degrees north. Instead, the most important event was the annual flooding of the Nile River in late spring, which irrigated all the farmland and made it fertile.

The Egyptians noticed that this event occurred very soon after the first predawn rising of the star Sirius, the brightest star in the sky, after it had been invisible for some months. This rising of Sirius became the cornerstone of the Egyptian year, which consisted of three seasons: Flood, Emergence of Land, and Low Water or Harvest.

THE KARNAK CLEPSYDRA
[AMENOPHIS III : CIRCA 1400 B.C]

these gates by observing a star as it crossed the meridian of the night sky. Thus the night came to be seen as divided into twelve parts, and the hieroglyph for these parts, these hours, was a star.

The division of the day into twelve hours, too, was simply by analogy with the night. The length of these hours varied according to the season, for in midwinter the time of darkness was longer than in midsummer, so a twelfth part of that time was also longer.

To establish equal hours, some other form of measurement was needed, and water clocks were designed and built from about 1500 B.C. "Star clocks" were also drawn up on elaborate charts, with specific stars marked as they transited the meridian, and they were used to identify each hour of the night throughout the year.

Babylonian Astronomy

The science in which the Babylonians achieved the greatest progress was astronomy, so much so that we still use some of their ideas and discoveries today. They studied meticulously the movements of the stars and planets, analyzing them mathematically so that they were able to predict their positions months or even years in advance. Their reasons for doing this were originally religious.

Unlike the Egyptians, the Babylonians became intensely interested in the planets (in ancient astronomy the Sun and Moon were also called planets) because they moved in complex paths among the fixed stars. The Babylonians could only explain this by supposing that the planets were divine beings. The stars and planets were called "the gods of the night." These astral gods were among the most important in Babylonian religion, and their movements were studied as omens of the future. Most of the astral omens concerned the welfare of the king:

"When on the fifth of the month Nissan the rising Sun appears like a red torch, white clouds rise from it, and the wind blows from the east, then there will be a solar eclipse on the twenty-eighth or twenty-ninth day of that month; the king will die that very month, and his son will ascend the throne."

Early Observatories

The king's scholars and scientific advisers had to study the heavens continually to establish what was normal and to observe anything that was abnormal and therefore a sign from the gods. The Babylonians soon made the crucial discovery that all astronomical events are periodic: risings and settings, eclipses and conjunctions, periods of visibility and invisibility—all these would recur in regular cycles.

The ziggurats served as temples and as observatories, and from their summits each night the astronomer-priests would record the movements of the heavenly bodies and analyze them into mathematical patterns. The heavenly bodies that most interested them were the Sun, the Moon, and

However, like most other peoples, the Egyptians also used the Moon to measure time, for the lunar month was a convenient way to subdivide the year. So these three seasons were divided into four lunar months, each named after a religious festival. This calendar suited farmers and priests, but the solar year cannot be divided evenly into equal months. In this case the rising of Sirius, although it occurred every 365 days, did not fall on the same date of the lunar month each year. This meant that months had to be adjusted now and then, which was complicated for any kind of record keeping.

So the Egyptians devised a second, more rigid calendar that ignored the Moon. For this calendar the predawn risings of 36 star groups were chosen, each of which rose just before the Sun for a period of ten days before being succeeded by the next. These stars groups, known as the "Decans," formed 36 artificial "months" of ten days each that together marked the Sun's passage through the sky for 360 days, and five extra days were added at the end of each year. This Egyptian calendar of 360+5 days formed the basis for all later Western civil calendars.

The Hours of the Day

The year, the month, and the day are natural divisions of time. But the division of the day into 24 hours is purely arbitrary, and we owe that, too, to the Egyptians. The Egyptians believed that each night the Sun god, Ra, traveled in a boat through the Underworld before emerging again at dawn. While in this Underworld, the god must pass through twelve gates guarded by demons. The priests were in the habit of marking the passage through

three stars are named as rising and becoming visible just before dawn—the "heliacal rising." The whole diagram was really a calendar, with each of the twelve months tied to an astronomical event. This shows that the Babylonians recognized that the Sun moves through the fixed stars over the course of the year, and that some stars are invisible for part of the year because they are in the sky during daylight hours.

Pictures in the Stars

By 800 B.C. the Babylonians had compiled star catalogs in which almost 100 bright stars were separately identified and grouped into constellations. A group of eighteen of these constellations were said to be "in the path of the Moon." The Moon's path is very close to that of Sun (it is inclined at around five degrees to it), and these eighteen constellations include the twelve zodiac constellations that we know today.

The Babylonians did not always see the same pictures in the stars that were later seen by the Greeks. What is important is that these star-groups became zones of the sky, landmarks in the map of the heavens. Some of the constellations that were the same as ours are Taurus the Bull, Gemini the Twins, and Scorpio and Leo. But Auriga the charioteer was seen as a sword, Aries was seen as a man, and Pisces the Fish as a bird.

When the Babylonians worked out the positions of the heavenly bodies, they did not plot them on a map with coordinates as we do. Instead, they calculated them in purely mathematical terms by working out units of movement against time, which they wrote down in columns of parallel figures. In this way a mathematical model of their motions emerged, which the skilled astronomer could use to predict future movements.

There were no geometric diagrams or celestial charts in Babylonia. They did not plot geometric paths for the heavenly bodies, and they do not seem to have produced theories about the physical structure of the cosmos. Their cosmology was, like that of the Egyptians, dominated by mythical beliefs. For example, they believed that the Sun-god spent the night under the earth, and they made pictures of him cutting his way up out of the ground at dawn. The Babylonian belief that the stars and planets were gods, and that they controlled human life, was the foundation of what later became the science of astrology.

Knowledge of the precise position of the planets was vital in assessing their power, but the Babylonians believed in that power because they were gods. The idea that the planets controlled human life as planets is a later Greek idea.

GREEK ASTRONOMY
Aristotle's Cosmos

His cosmology extended the ideas of Pythagoras and Plato, and gave them more physical reality. Aristotle was convinced that the Earth was a sphere (shown, for example, by the curved shadow of the Earth on the Moon seen during eclipses), and that the planets and stars revolved in spheres around the

Earth. The Earth was assumed to be the center of the universe, since all the heavenly bodies appeared to revolve around it.

The Sun, Moon, planets, and stars were thought to be carried on spheres made of some subtle, translucent, but definitely physical substance, and all rotated within one another as in a gear train.

The idea of empty space was alien to ancient thought, and Aristotle argued that each sphere abutted directly onto the next. He therefore recognized that intervening spheres would be necessary to neutralize or "unwind" the motions of the spheres.

The whole mechanism was driven in this way by the outermost sphere, which Aristotle called "the prime mover"; but what this original force really was, Aristotle was unable to say. It had to be the final source of all movement while remaining itself unmoved; otherwise, we have to imagine an infinite series of further spheres, each one moving the next.

Later, Christian and Islamic thinkers would identify the prime mover with God. But unlike Plato's "divine craftsman," Aristotle did not believe that the prime mover had created the universe; indeed, he considered that the universe had existed from eternity, following the principle that nothing could be created out of nothing. The heavens, he believed, were eternal and unchanging, and therefore could not be composed of the same four elements as the Earth, but of a fifth element, which he called the "ether."

Venus, which they called Ishtar. The first two determined the calendar, while Ishtar was one of the most powerful gods in Babylonian mythology. She controlled not only love but battles, storms, and fertility, and her position in the sky would indicate whether her influence was high or low at any given time:

"If on the fifteenth day of the month Shabatu Venus disappears in the west, remaining absent from the sky three days, and on the eighteenth day of the month she appears in the east, then catastrophe of kings: The weather god will bring rains, and the river god will bring waters from the earth: King will send greeting to king."

The Babylonians regarded the forces of the weather as under the power of the heavens. They did not know that the atmosphere was limited in its extent above the Earth, so they believed that clouds, storms, lightning, and so on all existed in the same realm as the stars.

By 1500 B.C. a series of "Venus tablets" had been compiled that gave the rising and setting times of Venus over many years, with omens relating to its position. They had recognized that the morning and evening star were one and the same.

A little later than the Venus tablets, but complete by 1000 B.C., was another series of tablets known by the title of "Three Stars Each." They are circular diagrams divided like a wheel with twelve spokes. In each segment

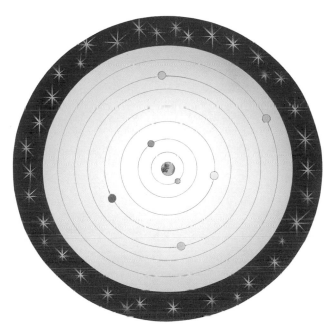

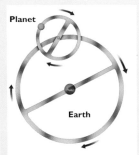

PTOLEMY'S THEORY OF EPICYCLES

Epicycle with axis on the deferent. This axis is rotating on the deferent, where the "deferent" is the main planetary orbit centered on the Earth.

The Achievements of Ptolemy

The central concept in Greek astronomy was that of the celestial spheres. The Earth was believed to lie at the center of a spherical cosmos whose outer shell carried the fixed stars. They were seen to rotate each night, apparently around a central polar point.

These stars did not move in relation to each other, so it was generally believed that they were luminous bodies that were somehow fixed on the celestial sphere, and therefore they all lay at the same distance from the Earth. Inside this outer sphere were further spheres, each carrying a planet, for the Greeks had identified several bodies in the heavens that moved independently from the other stars—the word "planet" meant "wanderer."

In classical astronomy the Sun and Moon were also called planets. These spheres all shared a common axis, which ran through the Earth too, and was thus regarded as the axis of the whole universe. This theory of the universe accounted well for the movements that people observed in the heavens. The entire cosmic structure was thought to be finite and self-contained: What, if anything, lay outside the starry shell was beyond human thought.

Greek skill in geometry permitted them to map and plot positions on the celestial sphere in the same way that latitude and longitude are used on Earth. Two base lines were used in this coordinate system. First, the altitude, or elevation, of a star or planet was measured from the ecliptic, the path of the Sun through the heavens (which had already been identified by the Babylonians).

Second, positions in azimuth, that is, around the sphere, corresponding to longitude on the Earth, were measured from a point in the constellation of Aries known as the vernal equinox. Technically it is the point where the ecliptic crosses the celestial equator. It stands at the beginning of the Zodiac and marks the beginning of the astronomical year and of spring on Earth. This coordinate system permitted any object in the sky to be given a precise location, and we know that celestial globes were made by Greek astronomers, marking the positions of many bright stars, although none has survived.

The constellations, most of them taken over from the Babylonians, were used to designate regions of sky. Ideas about the spherical cosmos had been developing since the time of Plato and Aristotle. The techniques of spherical astronomy emerged with the mathematicians of the second century B.C. such as Apollonius and Hipparchus. But the astronomer who summed up the achievements of classical astronomy and gave it its final form was Claudius Ptolemy, who worked in Alexandria between A.D. 130 and 170.

Ptolemy

In a great book, known by its later Arabic title, *The Almagest*, Ptolemy gave an account of the structure and movements of the heavens that was so complete that it remained scientific orthodoxy for fourteen hundred years. In the first place, Ptolemy presented a geometric model for the orbits of the Sun, Moon, and planets around the Earth.

Ptolemy and other Greek astronomers had realized that these orbits were much more complex than the simple circles suggested by Plato and Aristotle. Certainly the planets appear to circle the Earth each day, that is, they move in azimuth from east to west. But they appeared to have other motions too, for their positions also vary in altitude over a period of years.

This might suggest that the plane of their orbits was not fixed, but in the Greek model of the cosmos this must mean that the spheres bearing the planets were not fixed at the celestial pole, but were somehow oscillating. This seemed impossible, for it would mean abandoning the idea of the central axis of the cosmos.

Ptolemy's answer was to show that the planets possessed a double motion. A large sphere carried the planet on its daily journey around the Earth, but this sphere carried on itself a smaller sphere that was also revolving, but much more slowly, and this accounted for the smaller changes in a planet's position. The larger sphere was called the "deferent," that is, "the carrier," while the smaller sphere was called the "epicycle,' which means "the wheel lying on something else."

Taking observed positions of the planets over a number of years, Ptolemy carefully worked out in geometric terms the velocities and paths of all the planets with such precision that their positions at any time in the past or future could be pinpointed. In this view the whole mechanism of the cosmos was like a complex system of interlocking wheels moving in their unchanging paths. But this was a purely mathematical description: What kept the planets in their paths, and what was the power that moved them are subjects Ptolemy never discusses.

Ptolemy's second great achievement in *The Almagest* was to draw up a detailed catalog of over 1,000 bright stars, grouped into 48 constellations. He gives their positions in the celestial coordinate system and explains how they should be located on a celestial globe. Only a handful of the stars are named, for most of the star names that we now use originated with the later Arabic astronomers. Instead, Ptolemy gives them a number within each constellation and describes their position. Thus Aldebaran is described as "Number 14 of Taurus: the bright star of the Hyades, the reddish one in the southern eye of the Bull." This star catalog was to remain the fundamental guide to the heavens from the second century to the seventeenth.

We have seen that the Greeks regarded the universe as finite. In another of his works, The Planetary Hypothesis, Ptolemy sets out to calculate how big the universe is. He used calculations based on parallax, on the angular size of the Sun and Moon, and data drawn from eclipses to estimate the distances between the main celestial spheres.

The Moon, he said, was 160,000 miles from the Earth; the Sun, three million miles. He knew that the Sun was several times bigger than the Earth, while the Moon was considerably smaller. The distance from the Earth to the starry sphere, the outer shell of the cosmos, he estimated at 50 million miles; since the stars were conceived to be set in a single sphere at a fixed distance from the Earth, the diameter of Ptolemy's universe was thus 100 million

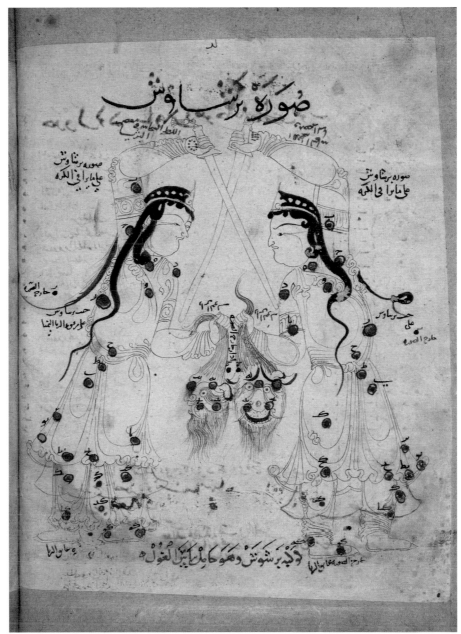

ARISTOTLE (384–322 B.C.)

- Philosopher, scientist, and physician.
- Born in Stagira, Macedonia.
- Went to Athens around 367 B.C. to join Plato's Academy first as a pupil, then as a teacher until Plato's death in 347 B.C.
- Moved to Assos, Asia Minor, then to Mytilene on Lesbos.
- Around 342 B.C. invited by Philip of Macedon to tutor Alexander his 13-year-old son and heir (Alexander the Great).
- 335 B.C. Returned to Athens to found the Peripatetic School at the Lyceum.
- Taught that the Earth is the center of the eternal universe. That everything below the orbit of the Moon is made up of earth, air, fire, and water and is subject to generation, destruction, qualitative change, and rectilinear motion; and that everything above the orbit of the Moon is made of ether and never changes but is subject to circular motion.
- Prolific writer on many subjects, including science, physics, metaphysics, ethics, logic, politics, and poetry.
- Believed that all material things could be analyzed using their matter and form, and that form shows their essence.
- Also believed that happiness is achieved by living in harmony with nature.

LEFT: The constellation of Perseus from a manuscript by al-Sufi. The figure is shown twice: once as seen from the Earth, and in mirror image, as on a celestial globe.

PTOLEMY (around 90–168 A.D.)

- Also called Claudius Ptolemaeus.
- Astronomer and geographer.
- Born somewhere in Greece.
- Spent much of his life working in the great library at Alexandria.
- Wrote *Geography*—a catalog of places, their latitude and longitude, description and details.
- Calculated the size of the Earth.
- Wrote *The Almagest*—a great catalog of the stars and everything known about them.
- Made many maps, including a map of the world.
- Wrote widely, especially about astronomy, the fixed musical scale, map-making, and chronology.
- The Ptolemaic System—orders the heavenly bodies (stars) into scientific form. Earth is the center of the universe with the stars revolving around it. In the ether above the atmosphere are seven concentric spheres each containing a planet and an eighth containing the fixed stars. It predicted the position of the stars to within one degree of accuracy.

miles—certainly a vast distance, but nevertheless still finite and within the bounds of comprehension.

Ptolemy's final work of celestial science was, perhaps surprisingly, a treatise on astrology called *Tetrabiblos* (meaning simply "Four Books"). In this work Ptolemy argues strongly that the stars and planets do influence events on Earth and human life. He believes this not because they are gods, as the Babylonians did, but because they emit forces other than heat and light that affect the Earth. He says it is undeniable than the Sun and Moon determine the weather, the seasons, and the tides, and he maintains that other, more subtle forces come from each heavenly body, and that they affect our minds and our bodies.

To Ptolemy this was not a matter of religion or of occult powers but part of nature's system—effects springing from physical causes. The relevance of the geometric astronomy of *The Almagest* to these beliefs is this: The astrologer has to calculate the precise positions of the stars and planets in order to assess their influence at any time, especially the moment of birth. Ptolemy's astronomy gave them the means to do so, and *The Almagest* became the fundamental technical tool of the scientific astrologer.

STAR GLOBES

The star globe represented the celestial sphere, with the Earth imagined to lie at its center. The stars were believed to be points of light set in this sphere, all lying at the same distance from the Earth. Therefore, when we look at a star globe, we are standing, so to speak, outside the celestial sphere, and all the constellation pictures will appear in the reverse of the way they appear when seen from the Earth. The pictures in al-Sufi's manuscripts show two versions of each constellation, one as seen from the Earth and the other in reverse as seen on a star globe. This suggests that these manuscripts may have been intended to be studied together with a celestial globe.

LEFT: Islamic celestial globe. Widespread among Islamic scientists from the eighth century, they were unknown in Europe before the fifteenth century.

Ptolemy's mathematical astronomy was immensely detailed and precise, and it was one of the summits of Greek science. Of course, it was based on a fallacy, for the structure of the cosmos is not as Ptolemy believed it to be. But by the standards of his day he gave a complete and precise mathematical description of what was seen in the heavens.

ISLAMIC ASTRONOMY: STARS AND OBSERVATORIES

Astronomy was central to Islamic science just it had been in ancient Greece, and for the same reasons. The majesty of the heavens and their eternal, unchanging nature were seen as powerful evidence that the universe was ruled by a divine intelligence. The movements of the heavenly bodies could be analyzed mathematically, and they also formed the subject matter of the science of astrology. Technically, Islamic astronomy was entirely developed from that of Ptolemy: The aim of analyzing celestial movements into regular geometric patterns became the fundamental aim of all astronomers. Sometimes they disagreed with the details of Ptolemy's theories, but their aim and their approach were always to refine and improve his system, never to replace it. The model of the cosmos accepted by Islamic scientists was that of Aristotle and Ptolemy, in which the stars and planets encircled the Earth in a system of concentric spheres.

The Islamic astronomers also developed a strong emphasis on the observation of the skies, motivated by the desire to improve the exactness of their cosmic model. Practical observation had long been traditional among the Arabic peoples: The nomadic tribes of the desert regions had built up a detailed knowledge of the stars, which were vital to them in gauging direction and time, in the same way that the stars were vital to mariners at sea.

Naming the Stars

It was the Persian astronomer al-Sufi (903–986) who set out to integrate traditional Arab star lore with Ptolemaic astronomy. This he achieved by revising Ptolemy's star catalog from *The Almagest*, giving improved coordinates for many of them, and by adding traditional Arabic names for several hundred stars. The Greeks, including Ptolemy, had given names to only a handful of stars, such as Sirius and Arcturus, while Ptolemy had identified stars by giving each a number within each constellation. Many of al-Sufi's star names were taken simply from the star's position in the constellation: Mintaka in Orion means simply "belt"; Markab in Perseus means "shoulder"; Rigel in Orion means "leg." The characteristic of the Arabic star names is the prefix "al," meaning "the." For example, Algol in Perseus means "the demon," while Aldebaran in Taurus means "the follower." These Arabic star names, first listed by al-Sufi, are still used by astronomers today.

Al-Sufi's great work of descriptive astronomy was called *The Book of the Constellations of the Fixed Stars*, and it was illustrated with vivid images of the constellation figures. This was another innovation over Ptolemy, whose star catalog was never illustrated in this way. The constellation figures also appeared on the celestial globes that were made by Islamic craftsmen from the eighth century onward. The use of star globes had a curious influence on the way the constellations were drawn in al-Sufi's manuscripts.

Astronomical Observatories

There were no star globes in Christian Europe at this time, nor were there any astronomical observatories such as those that existed in Islamic cities like Baghdad, Cairo in Egypt, Damascus in modern Syria, and Toledo in Spain. These observatories were not as sophisticated as modern ones, but their aims and methods were similar. They employed large sighting devices such as vertical or horizontal quadrants with which the angular distances between stars could be measured and the paths of the planets could be plotted. Such an observatory, and a long program of planned observations, lay behind many of the zijes. One of the most influential was that compiled by al-Zarqali in Toledo late in the eleventh century. Some astronomers in the Islamic kingdom of al-Andalus in Spain were uneasy about the complexities of Ptolemy's system and attempted to find simpler geometric models for the paths of the planets.

Another famous observatory was in Maragha in northern Persia, directed by the mathematician and astronomer al-Tusi. Al-Tusi (1201–1274) was present when Baghdad was conquered by the Khan Hulagu, the grandson of Ghengis Khan, and it was Hulagu who financed the new observatory, possibly motivated by his own interest in astrology. Hulagu's brother, the Khan Moengke, was the ruler of China, and he began the construction of a great observatory in Beijing. Al-Tusi supervised the observatory at Maragha from 1259 to 1274, using its excellent instruments and a staff of ten assistant astronomers to compile new tables of celestial positions and to refine Ptolemy's theories. Al-Tusi also wrote works on alchemy, mineralogy, and logic.

An even more ambitious observatory was founded in 1424 at Samarkand (in what is now Uzbekistan) by Ulugh Beg, a royal prince and grandson of the great conqueror Tamerlane. The centerpiece of this observatory was a huge arc cut one hundred feet deep into a hillside in the plane of the meridian. It was used to measure the altitudes of the stars; its remains can still be seen. With this, and with other instruments, Ulugh Beg compiled a new star catalog.

The death of this princely scientist was violent and mysterious: In 1941 his skeleton was identified by Russian archeologists, clothed in the robes of a martyr, and he had clearly been beheaded, but why and by whom has never been discovered.

Later, in the 1570s, an observatory was established at Istanbul that was richly equipped with sighting instruments. It is a striking fact that this observatory, and others like it in the Islamic world, never made the transition to the telescopic age. As late as the 1730s the ruler of Jaipur in India had constructed an elaborate observatory using massive naked-eye sighting devices similar to those employed by Ulugh Beg. This was long after the telescope had revolutionized astronomy in Europe, but intellectual contacts between the Islamic world and West were nonexistent, and no scientific revolution occurred in the East.

WESTERN MEDIEVAL ASTRONOMY AND COSMOLOGY

Astronomy in the Christian West was primarily descriptive, that is, its aim was to give a picture of the structure of the cosmos. But heaven was not only a physical place; it was spiritual realm too, the dwelling-place of God and his angels. For this reason astronomy interacted with religious and philosophical ideas probably more than any other science. Elementary astronomy had been preserved in certain texts such as that of Macrobius but by the late twelfth century the works of Ptolemy and Aristotle had been translated from Arabic into Latin, and the science became richer and more sophisticated.

There was universal agreement that the heavens must be understood in terms of spheres and circular motion, with the Earth at the center of the whole system. But exactly how many spheres there were, how they moved, and how they were interrelated were all matters of debate and disagreement. There was no final method of resolving these disagreements, no way of testing the different ideas. It was all speculation and theory.

One of the central problems was a disagreement between Plato and Aristotle on whether the universe was homogeneous—all composed of the same material—or not. Plato believed that it was, but Aristotle had divided the cosmos into two main regions. The region below the Moon, including the Earth, was composed of the four elements, where all was change, where movement occurred in all directions, and where things were born and decayed. Above the Moon, Aristotle argued, no change had ever been seen to occur, and the only form of motion was circular. Therefore this region must be composed of a different and eternal fifth element called the ether.

Celestial Spheres

Another puzzle was whether the universe was finite or not. Aristotle had taught that it was, and that neither space nor matter existed beyond the last sphere. But in the religious culture of the Christian Middle Ages this seemed to place a limit on God's power; indeed, this was one of the Aristotelian ideas that had been included in the condemnation of 1277. Many theologians argued that God was capable of creating another universe, but in that case space must exist outside this one in which this could happen. The possibility was even suggested that the universe might be infinite, since God could surely create multiple universes if he chose.

Within our cosmos how many celestial spheres were there? All the stars were believed to be set as points of light on one sphere that revolved at a fixed distance from the Earth. Then there were the five planets known since

LEFT: The spheres of the medieval cosmos. This fine representation of the universe of the time is on display in Pisa, Italy.

BELOW RIGHT: Pawnee buckskin chart of the night sky. It was used in divination and recalls the Pawnee reverence for the stars and planets.

PAGE 22: Twentieth-century bark painting showing a Wandjina. The Wandjina are a group of ancestral beings from the sea and sky who bring rain and control the fertility of the animals and the land. They are shown in bark paintings and rock art throughout the Kimberley region of Australia.

PAGE 23: Part of the bronze astronomical observatory constructed by Ferdinand Verbiest for the Chinese Emperor Qing Kangxi between 1669 and 1673.

ancient times, Mercury, Venus, Mars, Jupiter, and Saturn, plus the Sun and Moon. All these bodies moved independently of the fixed stars; therefore there must be eight spheres. But most astronomers believed that these spheres moved like a gear train, and that their motion arose from another sphere outside the whole system—the primum mobile, or first mover.

The Bible also spoke of "the waters above the Earth," and to take account of this, a further sphere of frozen, crystalline water was added; and then, above that, a final sphere of heavenly fire, the empyrean, was added, making ten in all. However, it seemed probable, as Aristotle had suggested, that the planetary spheres could not simply turn against each other. Since there was no empty space between them, they would need intermediate spheres to "unwind" or neutralize the adjacent motion. And then, as we have seen in the Ptolemaic system, each planet required an extra epicycle to account for their periodic changes of position in addition to their main daily cycle around the Earth.

The epicycle spheres were usually considered to be embodied in the thickness of the main deferent sphere. Taking account of these possibilities, some theorists produced schemes in which there were fifty or more celestial spheres, all revolving around the Earth in complex interlocking movements. Each sphere touched the next, with no empty space intervening. What these spheres actually were, nobody knew. They must be invisible or translucent, perhaps made of some form of glass, perhaps fluid. It was believed to be possible to estimate the total size of the cosmos, just as Ptolemy had by using parallax calculations. The radius of the entire cosmic system, from the Earth to the abode of God and his angels, was usually considered to be on the order of fifty million miles.

The Unmoved Mover

The final but shadowy question behind all medieval cosmology was the power that moved these spheres. Each sphere turned the next, but somewhere the sequence must have a beginning. The idea of an unmoved mover was easily identified with God. However, the problem was that Aristotle had suggested not one single prime mover but several, one for each of the celestial spheres. In the Christian system it was clearly impossible to believe that multiple divinities inhabited the heavens, so perhaps these celestial forces were the angels; and indeed, in many medieval illustrations angels are shown turning the heavenly spheres by means of crank handles.

A few very original scholars produced a completely different theory of celestial motion. The Paris teacher Jean Buridan (1215–1279) suggested that God had given impetus to the whole system at the moment of creation, and that it was still turning freely as a result. This idea was very daring and controversial, just as the debate about primary and secondary causes had been. It implied that God's power was limited to the time of creation only, and that thereafter nature functioned as a self-sufficient system. This was not acceptable to many theologians.

Astronomical Data

When Western scholars awoke to the possibilities of mathematical astronomy after the twelfth century, they had no body of observational data to build on. Therefore they borrowed the mathematical tables compiled by Islamic scientists, especially those produced in Toledo in the eleventh century and another set drawn up in Spain after the Christian reconquest, the Alfonsine Tables—so named because they were dedicated to King Alfonso the Wise in 1252. The Toledan and Alfonsine Tables provided European scholars with all their astronomical data for several hundred years, until they developed the skills to draw up accurate tables of their own. No observatories on the Islamic model were established in western Europe, although the use of the astrolabe became wide-spread by the fourteenth century.

In the medieval view the cosmos was a subtle, complex, but finite mechanism. It was the self-contained realm over which God ruled, and of which our world was one central part. Although its details could be endlessly argued over, its general structure was comprehensible to all, and it dominated the European mind until the Renaissance.

The Meeting place of Science and Religion

It is this vision that lies behind Dante's great poem *The Divine Comedy*. After visiting hell and purgatory, in the final part of the poem, *Paradiso*, Dante is taken up from the Earth on a journey through the ten celestial spheres, and he describes the character of each. In medieval thought, as with Plato, the heavenly spheres were the dwelling places of the human souls, who have been placed in the sphere most fitting their lives. In that of the Sun, the source of light, Dante meets St. Francis of Assisi and St. Thomas Aquinas. In the sphere of Jupiter he meets great rulers such as King David and the Emperor Constantine. Among the stars he meets the souls of the Apostles; and finally, in the tenth sphere of fire, beyond the material universe, he comes face to face with God. This vision shows astronomy as the meeting place of science and religion.

Astronomy in Africa, North America, and Australia

Evidence of the study of astronomy is found in cultures in every part of the world. The positions of the heavenly bodies were used to gauge time and direction, and they were very often seen as embodiments of cosmic powers. Until fairly recently it was believed that astronomy flourished only in the major civilizations of the Near East, China, or India; but it is now recognized that many cultures in Africa, pre-Columbian America, and the Pacific also developed a high degree of astronomical knowledge.

Seven Stars

The Borana people of southern Ethiopia and northern Kenya used a unique calendar system based on the conjunction of seven selected stars with the various phases of the Moon. Near Lake Turkana in Kenya an array of stone pillars has been discovered that were used to make the necessary sightings. Twelve months were designated, each beginning when the new crescent Moon rose in the same position on the horizon as a bright star, such as Aldebaran in Taurus, the Pleiades, those of Orion's belt, and so on. It has been estimated that the Turkana site dates from around 300 B.C., and the star positions would have been correct for that date. The Borana calendar represents a unique approach to astronomy based on reasoning and on detailed observations that must have extended over many decades, perhaps over generations.

Other sites that used stone structures in astronomical alignments have been found in sub-Saharan Africa. The Great Zimbabwe site, inhabited probably from A.D. 400 onward, has a number of markers built into it, including rock patterns that mark the rising Sun at the summer solstice, a passageway giving a view of the Milky Way overhead and also at the summer solstice, and two large stones placed to form a slit through which the Sun rose precisely at the two equinoxes of the year. It seems certain that these societies maintained dedicated astronomers whose task it was to make the necessary observations.

Mysterious Knowledge

Perhaps the best known, and most mysterious, example of astronomical knowledge in Africa is the case of the Dogon tribe of Mali. In the 1930s anthropologists studying this tribe discovered that the Dogon had a special interest in the star Sirius, which is the brightest single star in the heavens. The Dogon believed that Sirius had an invisible companion, which was smaller and heavier than Sirius, and whose power both helped grain ripen and kept the other stars in their places. Sirius and its invisible companion were said to orbit around each other every fifty years.

Why is this so mysterious? The answer is that in the 1860s European astronomers had discovered than Sirius was a binary star, that is, it has a twin star, and that the two mutually orbit each other. Sirius's twin was found to be much smaller and denser than Sirius—it was the kind of star later known as a white dwarf—and is visible only through very powerful telescopes. The orbital period of Sirius and its companion was found to be fifty years. The Dogon have rituals and ceremonies every fifty years to mark the completion of such a cycle.

How is it possible for this tribe in Mali to have been in possession of highly specialized astronomical facts that were known to Western astronomers only from the late nineteenth century? The rational explanation is that the Dogon somehow learned the scientific truth about Sirius and embodied it into their mythology. This is theoretically possible, given European contacts with Mali in the nineteenth century. At the other extreme is the suggestion that alien beings from Sirius visited the Dogon in the remote past and taught them these secrets. Whatever the true explanation, this case is a striking reminder of how little we know, and how much may still remain to be discovered about the ancient astronomy of non-Western peoples.

American Astronomy

Among native North American peoples astronomy was equally widespread and important. One peculiarity of this region was the attention paid to the Pleiades, the distinctive group of seven stars in the constellation Taurus. As early as 1524 the explorer Verrazano noted that the indians of Rhode Island were guided in their spring planting of crops by the rising of the Pleiades, while Pleiades lore is found in all but 12 of the 58 native cultural territories.

As in Africa, evidence of astronomical study has been found embodied in sites all over North America. In Chaco Canyon, New

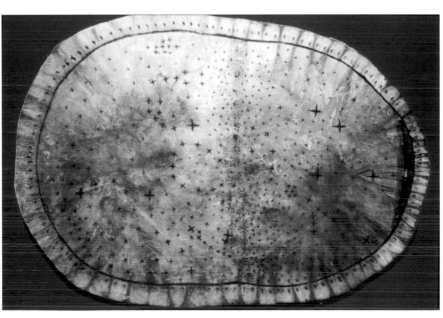

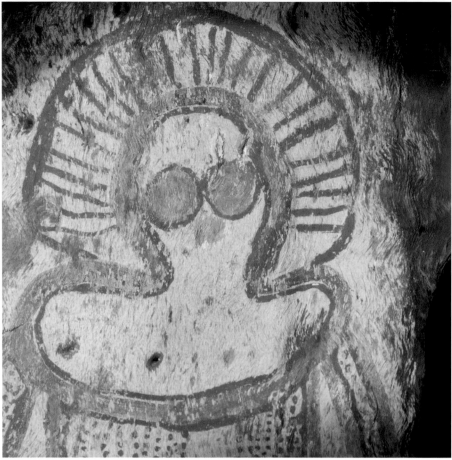

purpose. A ring of stones 87 feet in diameter with 28 spokes, it seems to be aligned northeast to southwest, on the sunrise at the midsummer solstice and the sunset at the midwinter solstice. The age of the Medicine Wheel and the identity of the people who built it are both uncertain, but it resembles other smaller features, especially in southern Canada, that are known to be thousands of years old. The importance of the Sun in the rituals and religions of the Plains Indians is well known from the Sun Dance, in which pain and fasting while gazing at the Sun were thought to purify and strengthen the participant. Thus, in addition to calendar functions, there may have been religious motives behind these structures, just as there may have been at Stonehenge. The great site of Cahokia in the Mississippi Valley contained a very large Sun circle 410 feet in diameter, with perimeter posts aligned on the solstices and equinoxes.

But other features in the sky apart from Sun were also important. In Nebraska the Omaha people erected a Sacred Pole to symbolize tribal stability. This pole was oriented on the North Celestial Pole, the point around which the heavens appear to revolve. The Chumash of southern California saw the Milky Way as the route to the sky followed by the souls of the dead. The importance of astronomical observations to these peoples was summed up in the memoirs of a southern California chief:

"When the Sun swung to the north, and the Moon showed quartered by day overhead, they knew by the signs of the Sun and Moon when the seeds of certain plants were ripe, and they got ready to go away and gather the harvest. Every plant that grew, the nesting time of all birds, the time of the young eagles, everything they learned by the signs of the Sun and the Moon."

Rich Australian Tradition

We know that the aboriginal tribes of Australia had an equally rich tradition of using celestial markers to tell them when foods were ripening or when animals were breeding. We know too that they embodied their astronomical knowledge in many celestial myths in which the Sun, stars, and planets were seen as cosmic powers governing time, fertility, and harmony in the universe. We must always remember that these people spent much of their lives in real darkness—the kind of darkness that modern industrial society has banished. They must have spent countless hours watching the heavenly bodies, noting the precise patterns in which they moved, and devising myths and legends to explain what they were seeing in the sky.

CHINESE ASTRONOMY: CELESTIAL SPHERES

Astronomy developed at an early stage in China, and as in Mesopotamia, it was associated with astrology. From around 1200 B.C. omens mention certain stars by name and give warnings concerning eclipses. These omens relate entirely to the emperor or his enemies, not to the common people. The similarity with the Babylonian system of divination raises the question of possible influence from Babylonia, although no links between the two cultures have ever been established.

There was an important distinction between Babylonian and Chinese astrology however. In the former the gods used the heavenly bodies to give warnings to the king; but in China the emperor might cause unusual events in the heavens. This was because the emperor was regarded as the son of heaven, as contributing to the harmony of the universe. If the emperor failed in his duty or ruled unjustly, that would be seen in eclipses, comets, supernovas, and so on. The theory of yin and yang also played its part. From high summer to midwinter yang, the male power of light, was ebbing. At the midwinter solstice the emperor performed the most important ritual of year, after which yang began once again to increase.

Zoning the Sky

Chinese astronomers did not recognize the Mesopotamian zodiac; in fact, they did not use the path of the Sun at all when dividing the sky into zones. Instead, there were several different ways in which they zoned the sky. First, they designated 28 "Lunar Mansions," that is, groups of stars in which the Moon was to be found on successive nights during its monthly circuit around the Earth. This system was in place by 1000 B.C. at the latest. Unlike the zodiac signs, the lunar mansions were of different widths, but they each covered an average of 13 degrees of celestial longitude. Second, in addition to these 28 sky zones the Chinese designated around 280 small constellations, some containing only four or five stars. These constellations were not the same as the Western ones, with a few exceptions: Orion and the "Big Dipper" section of the Great Bear were patterns that the Chinese also

Mexico, are a number of pictographs, including one of a large star having 23 rays, together with a crescent Moon. Some experts have argued that this represents the supernova explosion of July 1054, which we know from Chinese records, and which first rose with the waning crescent Moon and remained visible even in the daytime for 23 days. Also in Chaco Canyon a spiral rock carving interacts with shafts of midday sunlight at the solstices and equinoxes in a way that shows precise knowledge of the timing of those events; this feature has become known as the "Anasazi Sun-Dagger."

On a much larger scale is the Bighorn Medicine Wheel in Wyoming, which has been compared to Stonehenge in England in its alignment and

recognized. And finally, there was a division of the sky into "Five Palaces." The most important of them contained the stars around the north celestial pole, which never set, while the others were in bands south of that group.

Celestial Sphere

How then did the Chinese view the structure of the cosmos? There were no great cosmological systems in the Western style comparable to those of Plato, Aristotle, or Ptolemy. In the earlier phase, around the fourth or fifth centuries B.C., there were a number of simpler word pictures of the universe, which suggest that the heavens were seen as a semispherical bowl place over the semispherical Earth. Some five or six centuries later, in the first century A.D., astronomers began speaking of the heavens as forming a complete sphere in a way that is remarkably similar to the Greek model. For reference purposes they divided the sphere into 365 units—the distance traveled by the Sun in a day, and they, too, produced a model of the celestial sphere, with poles, tropics, equator, and equinox points. They also used star catalogs that listed almost 1,500 stars, giving them reference positions on the celestial sphere, just as Ptolemy and the Greek astronomers did.

Whether this system was devised independently, or whether the Greek model was somehow transmitted to China, no one has ever been able to answer. It seems unlikely that the Chinese were aware of Greek astronomy in any detail, for they never went on to produce mathematical models of the paths of the Sun, Moon, and planets. They did try to explain the complex paths of the planets, but by devising geometric patterns that were not given mathematical form. For example, in the eleventh century A.D. Shen Kua explained the retrograde motion of Mars or Jupiter by suggesting this "willow-leaf" model: that the planet left its circular path from time to time to follow a very sharp ellipse, then returned to its circle. This would have been impossible in Greek astronomy, where all heavenly bodies were thought to move in uniform circular motion.

There was nothing in Chinese astronomy comparable to the interlocking-spheres model of the Greeks. On the other hand, the Chinese were much freer to speculate about the ultimate structure of the universe, and some astronomers even suggested that it was in fact infinite, with the Earth and stars floating freely in it. They were not tied together in any systematic way but drifting like leaves on a stream. Chinese scientists did use the spherical cosmic model to estimate the circumference of the Earth, just as the Greeks did. In A.D. 725 shadows were measured simultaneously on the same meridian, although the Chinese result, at just over 32,000 miles, was around twenty-five percent too big.

Detailed Records

Chinese attention was fixed on observing and recording data, rather than on system building. One of the results of this is that the Chinese have the longest and most detailed record of astronomical events from any

IOHANNES d. REGIO MONTE dictus
alias MVLLERVS.

LEFT: Regiomontanus, who revised and updated the data in Ptolemy's *Almagest*, and opened a new era in astronomy.

RIGHT: Copernicus's drawing of the Sun-centered cosmos illustrates his *De revolutionibus orbium coelestium.*

PAGE 28: Copernicus's revolutionary ideas raised new questions about the stars: What were they, where were they, how was man to understand his place in this new universe? This is the wonderful ceiling fresco of the constellations in Caprarola, Italy.

PAGE 29: The scientific and intellectual revolution of the late Renaissance had its roots in astronomy. This picture is entitled "The Comet of 1532." It would not be until the seventeenth century and the work of Halley that comets would be explained. Even Galileo thought that they were optical illusions.

NICOLAUS COPERNICUS (1473–1543)

- Originally named Nikolaj Kopernik.
- Founder of modern astronomy.
- Born Torun, Prussia (now Poland).
- 1491–94 Studied mathematics and optics at Cracow University.
- 1496 Went to Italy to study canon law and attended astronomy lectures at Bologna University.
- 1497 Nominated a canon at Frombork cathedral—but never took holy orders.
- 1501–05 Learned medicine at Padua. While there, in 1503, made a doctor of canon law by the University of Ferrara.
- Returned to Frombork, Poland, as an administrator and the bishop of Ermeland's (who was also his uncle) medical advisor.
- Deeply troubled by Ptolemy's assertion that the Earth was the center of the universe.
- 1512 Started a mathematical explanation of the Sun being the center of the universe, but unwilling to publish for fear of the church establishment.
- 1543 By now old and ill, he was reluctantly persuaded by his pupil Rheticus to finally publish his theories in *De Revolutionibus Orbium Coelestium* ("On the Revolutions of the Heavenly Spheres"). He dedicated the work to Pope Paul III. Copernicus received his copy on his deathbed.
- The book scandalized the establishment and was immediately banned by the Catholic church and placed on the list of forbidden books, where it remained until 1835.

civilization. Ptolemy and the Greeks were far less interested in unusual events like supernovas or eclipses because they did not appear to obey astronomical laws. The Chinese, on the other hand, eagerly recorded all such events. Sunspots were studied and described from the first century B.C. onward, being observed through smoked glass or jade. Solar eclipses were closely observed from the eighth century B.C., to such effect that Chinese astronomers were soon able to predict their occurrence.

Comet records extending over almost two thousand years make it clear that the Chinese had identified the periodic return of Halley's comet. Supernovas—exploding stars—were recorded in 1006, 1054, and 1572. Until the last date they were seen in the West as miraculous and were not recorded by astronomers, for the accepted doctrine was that the heavens were eternal and unchanging. The 1054 supernova is still visible and is now known as the Crab Nebula, while the 1572 supernova was the one that Tycho Brahe regarded as a new star, and that convinced him that change did occur in the heavens.

Watching the Skies

The Chinese produced optical sighting instruments such as the quadrant and the armillary, and several well-equipped observatories were built to house them. They were state institutions, maintained in order to calculate calendars and to feed the emperor with astrological data. A rich tradition of star maps developed, with individual stars, the constellations, and the lunar mansions charted with celestial coordinates. Perhaps the most remarkable instrument the Chinese ever built was the large waterclock that turned into a model of the celestial sphere. It was the work of the scientist Su Song in the eleventh century A.D., and it included an escapement, that is, a retarding mechanism that ensures that the model turned at the same speed at which the Earth rotates.

Chinese astronomy was rich and detailed, and it reflected some of the philosophical principles of its parent culture, just as Western astronomy did. It was a science strong in observation and data gathering, but less interested than the West in building systematic explanations of what it saw in the heavens.

RENAISSANCE ASTRONOMY

Before Copernicus

In the fifteenth century astronomy involved very little empirical or observational work. Its main function was twofold: to study and clarify the structure of the cosmic spheres, as described in the classical theory of Ptolemy, and to underpin the practice of astrology. For both of these purposes the astronomer's work consisted of mathematical calculation of celestial positions. Optical sighting devices did exist, but most astronomers and astrologers relied on precalculated tables of celestial positions and mathematical formulas. The most widely used of these tables were the so-called "Alfonsine Tables," which had been compiled in the middle of the thirteenth century. Thus the astronomer of the thirteenth century was distanced from the direct study of the heavens by the rigid theory of cosmic structure that had been inherited from the remote past and by the use of precomputed data.

This situation began to change in the years 1460–90 with the innovative work of two German astronomers, Georg Peurbach (1423–61) and Johannes Muller (1436–67), who is always known by his Latin name of Regiomontanus. These two men decided to undertake a thorough revision of Ptolemy's great text *The Almagest*, first by taking a small number of very careful optical sightings of celestial positions, and then by working through his calculations. *The Almagest* was the great authority in medieval astronomy, but it is a long and technically difficult work, and most people knew it only in shortened versions, while no Western astronomers had ever rechecked its accuracy in this way. Peurbach and Regiomontanus then intended to compare their results with the data given in the Alfonsine Tables.

Peurbach died when the task was less than half finished, but Regiomontanus carried it through to completion by the year 1463. The resulting book was published some years later as an *Epitome of Ptolemy's Almagest*, and it was of enormous importance. Regiomontanus had revealed so many errors in the Alfonsine Tables that he exclaimed, "The common astronomers of our age are like credulous women, receiving as something divine and immutable whatever they find in books, for they make no effort themselves to find the truth." The motions of all the planets were found to be incorrectly plotted in the tables, so that their appearances, conjunctions, eclipses, and so on all might take place many degrees away from the place predicted and days before or after the time predicted.

Belief Without Question

The interesting thing about Regiomontanus's work, however, is that he did not question the mathematical models that he found in Ptolemy. He assumed they were correct, while the observations of positions underlying the Alfonsine Tables were wrong. The reform of astronomy, Regiomontanus argued, must come from far more accurate observations from which the paths of the planets could be worked out precisely, still using Ptolemy's theories. Regiomontanus set in train a process of reform in astronomy whose end would have astounded him: He revealed discrepancies between observed reality and the standard texts that would only be explained after the revolution wrought by Copernicus, Tycho Brahe, and Kepler. Copernicus himself studied Regiomontanus's work and was struck by the problems it revealed, but he concluded that it was the Ptolemaic framework itself that was at fault.

Regiomontanus settled in Nuremberg and established a printing press with which he intended to disseminate scientific works; but he, too, died prematurely. He was a transitional figure who stood on the threshold of a new science. Although he did not dream of questioning Ptolemy's cosmic model, he introduced an important new note of empiricism into astronomy, looking at the heavens with fresh eyes.

Copernicus and His Revolution

In the year 1543—many decades after the invention of printing, after the discovery of the New World, and after the rebirth of art—there occurred an intellectual event that changed forever humans' understanding of themselves and their world: Copernicus published his theory that the Earth was not the center of the universe, but was a planet orbiting the Sun.

The idea that the Earth was the central point of the cosmos had seemed to follow from the evidence of our own eyes, and it had been universally accepted by all thinking people. The rejection of that idea was one of the great landmarks in humankind's intellectual history, whose profound implications required many years before they were fully understood.

Nicholas Copernicus (1473–1543) was a mathematical astronomer born in the town of Torun, which stood then on the borders between

Germany and Poland. His original name was Nikolaj Kopernik, but he Latinized his name, as did many scholars at this time. Copernicus was a churchman by profession and an astronomer only by inclination, and his role in the revolution that bears his name is not as simple as is sometimes supposed.

Elegant Theory

The natural model of a scientific revolution is to imagine scientists working within an agreed framework. They are confronted with new data that proves impossible to reconcile with that framework, so they are forced into a new understanding of their subject. Yet nothing like this happened in the case of Copernicus. New facts, new astronomical observations, new evidence—all these are absent from his work. He was a student of books rather than of nature, and observation of the sky was not the basis of his new theory. Instead, he performed a highly original thought experiment: He devised a new geometric model that accounted for the movements that we see in the heavens in a simpler, more elegant way than the classical theory of Ptolemy had.

Copernicus did not leave a detailed account of the way his thinking developed, but he said enough for us to know that he was dissatisfied

easily explained by supposing that they revolved around the Sun and nearer to it than the Earth. But Copernicus still accepted the physical reality of the celestial spheres: Therefore it was impossible that these two planets should orbit the Sun while the Sun and the other still orbited the Earth, for the spheres would have to intersect. Only by placing the Sun at the center of the system and the Earth as a planet in motion around it could all movements of the planets as we see them be explained.

Copernicus still had to account for certain irregularities in the planets' paths—which we know are caused by the fact that their orbits are not perfectly circular. Copernicus did this in the traditional way by inventing more epicycles, showing that he was still approaching the problem in a purely mathematical way. Galileo would later point out that the path ascribed by Copernicus to Venus could not explain that planet's constant brilliance, which showed that personal observation of the sky played no part in Copernicus's method. The Copernican revolution was a purely conceptual one. It was not a discovery, for he was never able to offer any evidence that it was true. Other mathematicians could only consider his model and try to assess how it accounted for what we see in the sky. And in any such assessment the idea that the Earth was spinning through space seemed to contradict our common experience, which tells us that the Earth is at rest.

Reluctance to Publish

Copernicus's theory was essentially complete by the year 1510, and within a year or two of that date he had circulated it in manuscript among a few friends. Long years passed, during which Copernicus must have pondered the implications of his ideas, but he did not publish them. There is one very revealing detail about his thought processes during these years. In his original manuscript, which still survives, Copernicus referred to the ancient Greek astronomer Aristarchus, who had suggested the same idea of a Sun-centered universe long before. But Aristarchus had narrowly escaped being put on trial for his outrageous and irreligious views, and Copernicus canceled this reference when his book was finally published. There seems little doubt that he hesitated to publicize his idea because he was aware how controversial it would be, and in particular he must have foreseen trouble with the church authorities.

We cannot be certain that he would ever have published it at all had not his hand been forced by one of his friends, George Rheticus, who in 1540 printed his own brief account of the new theory. Copernicus naturally wanted to give his theory in full in his own way, so he set to work to prepare the long-withheld manuscript for the press. The book was printed in Nuremberg in 1543, with the title *De revolutionibus orbium coelestium* ("On the Revolutions of the Heavenly Spheres"). Whether the famous story is true that a copy was placed in Copernicus's hands on May 25 of that year, as he lay dying, we can never be sure. We only know that he died in almost complete obscurity, presumably unaware that his name would be linked with one of the greatest intellectual revolutions in history.

with the Ptolemaic theory of the heavens from an early stage in his career. Its complex system of epicycles and eccentrics seemed to Copernicus to be impossible to reconcile with the movements of real physical objects in space. These were doubts that many astronomers had experienced over the centuries, but none of them had ever seen the correct way out. It occurred to Copernicus, however, that the movement may be real or apparent: We can easily see how apparent movement may be caused by the observer's own movement—for example, if we sit in a spinning chair, the room appears to revolve around us. This gave Copernicus his vital clue, and he later wrote:

"A seeming change of place may arise from the motion of either the object or the observer . . . . If then some motion of the Earth be possible, the same will be reflected in external bodies, which must seem to move in the opposite sense . . . . I began to think of the mobility of the Earth; and though the idea seemed absurd . . . . I considered that I might find, by assuming some motion in the Earth, sounder explanations for the revolutions of the celestial spheres."

And so it proved, for looking first at the movements of Mercury and Venus, Copernicus saw that their constant proximity to the Sun was much more

The Response to Copernicus

The Copernican revolution was one of the great turning points in intellectual history, but it is important to understand that it was not an event like the French Revolution, which changed the world in a few days. Copernicus's new theory of the solar system appeared in print in the early summer of 1543, but more than half a century was to pass before it was widely accepted by scientists and scholars, and even longer before its implications were fully worked out.

Why was this? The first objection to Copernicanism was the most obvious one, that it entailed belief in an Earth that was whirling through space and rotating on its axis. How could this be true, contemporaries asked, when their senses told them that the Earth was motionless? If an object such as a stone is thrown up into the air, it does not land miles away as the surface of the Earth rushes by beneath it.

As one philosopher wrote: "No one in his senses will ever think that the Earth, heavy and unwieldy from its own weight and mass, staggers up and down around its own center or that of the Sun, for at the slightest jar of the Earth, we would see cities and fortresses, towns and mountains thrown down."

Only gradually did it come to be understood that this does not happen because everything on the Earth is sharing the Earth's motion. The stone thrown vertically up into the air is moving with the Earth and therefore returns to the same point that it left. The stone has moved in space in other ways, too, but this would not be apparent to any observer standing on the spot who also shares the Earth's motion. It was Galileo who made this clear much later through his analogy of an object dropped from the mast of a moving ship: Everyone would expect the object to land at the foot of the mast, not some yards away as the ship moves. So this objection, although at first so powerful, was capable of a straightforward solution.

Infinite Universe

The second problem was more subtle and more profound, and it concerned the scale of the universe. The classical universe of Ptolemy was very large, but it was finite; it was closed by the outermost sphere, which carried the stars. The stars were said to be fixed because their positions on that sphere never changed in relation to each other. But if the Earth was indeed moving in an orbit around the Sun, then the Earth's position would change by many of millions of miles in the course of the year. One of the effects of this movement should be that the positions of the stars should change enormously as viewed from the Earth. This would be an effect of parallax. But Copernicus noticed, and others after him noticed too, that no such parallax effect occurred.

Even Tycho Brahe, the most dedicated astronomical observer of his age, searched in vain for any such parallax changes in stellar positions. The only possible explanation for this was that the stars were so far away that no parallax shift could be detected. This, in turn, implied that the stars were far,

far further from the Earth than had ever been imagined, that the scale of the universe was much vaster than had been supposed. Moreover, there was no longer any reason to believe that the Earth or the Sun lay at the center of the universe. Therefore the human world was displaced from being the focus of the whole creation to being a random point in a huge and uncharted universe.

"The heavens," concluded Copernicus, "are immense and present the appearance of an infinite magnitude." The crucial word here is "infinite." The idea of an infinite universe was to have an immense impact, especially on religious thought. Why should God create an infinite universe? If the great goal of history was the salvation of humankind, why should that take place in an infinite space?

The idea that the Earth is in motion and the idea of an infinite universe both directly contradicted classical Aristotelian science, and that was the real reason for resistance to Copernicanism. All the scholars and philosophers in

Europe had been trained to accept the picture of the universe as a finite system enclosed within a nest of spheres, a mechanism ruled by God. This was the universe of the medieval church, of St. Thomas Aquinas and Dante, and it was comprehensible and rational, while the new universe of Copernicanism was neither.

Aristotle had taught that falling objects were seeking the center of the universe, which was also the center of the Earth, and that all natural motion is to be explained in this way except the spherical motion of the heavens, where other laws exist. But if the Earth were not the center of the universe, why should heavy objects fall downward at all? In the Aristotelian universe the heavier elements—earth and water—gather at the center, while the lighter—air and fire—rise upward. But again, if the Earth is not the center, this could not be true. Copernicus suggested instead a principle of cohesion in which all heavy matter would gather together into a sphere, although this

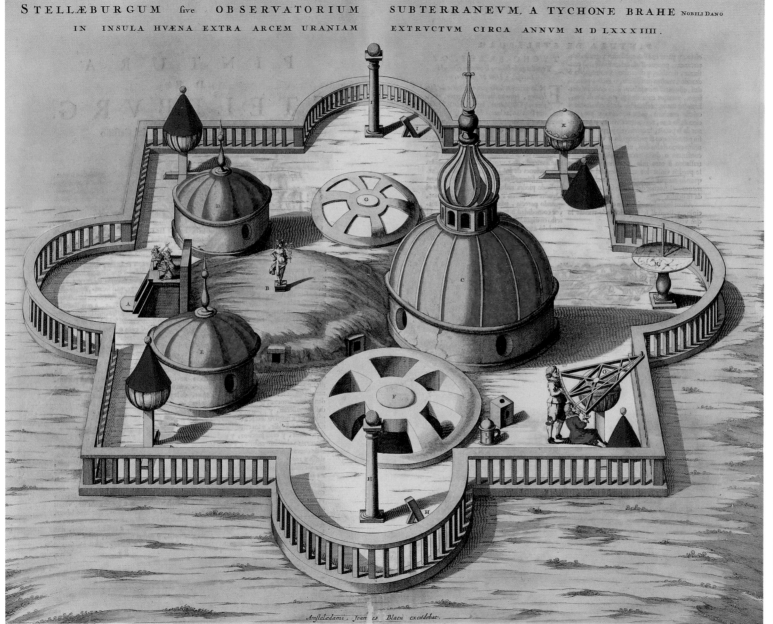

STELLÆBURGUM five OBSERVATORIUM SUBTERRANEVM, A TYCHONE BRAHE NOBILI DANO

IN INSULA HVÆNA EXTRA ARCEM URANIAM EXTRVCTVM CIRCA ANNVM M D LXXXIIII.

Amsteledami. Joannes Blaeu excidebat.

TYCHO BRAHE (1546–1601)

- Astronomer who accurately cataloged the stars before the discovery of telescopes.
- Born in Knudstrup, Sweden, then under Danish rule.
- In 1565 lost most of his nose in a duel and wore a silver nose for the rest of his life.
- Studied math and astronomy at University of Copenhagen and then at Leipzig, Wittenberg, Rostock, and Augsburg.
- In 1572 made his name by noticing a new star in Cassiopeia; the supernova is known as Tycho's star.
- 1576 Founded Uraniborg Observatory on the island of Ven, (formerly Hven).
- For 20 years he cataloged and measured the positions of 777 stars until thrown out of Denmark in 1596.
- Finally, after three years traveling, settled near Prague, where he was assisted with his observations by Johannes Kepler .

LEFT: Uraniborg, Tycho's unique observatory on the island of Ven; parts of the roofs were removable for observation. The lower photograph shows the subsidiary observatory on Ven.

RIGHT: In spite of Kepler's pioneering work in astronomy, he believed deeply in astrology convinced the heavens played a role in human behavior. Astrologists believed that mankind's mind and body were in harmony with the cosmos, as shown in *Ars Magna Lucis et Umbra*, published in 1646 by Athanasius Kircher (1602–80).

JOHANNES KEPLER (1571–1630)

- Astronomer.
- Born in Weilderstadt, Württemberg, Germany.
- Educated at the University of Tübingen.
- Professor of mathematics at Graz from 1594.
- Astrologer to Duke Albrecht of Wallenstein.
- *Mysterium Cosmographium*, ("The Mysterious Cosmos") an explanation of the geometric relationship between the Sun and the six planets, published 1596.
- Invited to Prague to help Tycho Brahe in 1600; on Tycho's death in 1601 succeeded him as imperial mathematician to Emperor Rudolf II.
- Published *Tabulae Rudolphinae* ("Rudolf's Tables") explaining Kepler's first, second, and third laws of the planets, plus a catalog of 1,005 stars in 1627.

need not be the center of the universe. But if the Earth is a planet, might not all the other planets be earths? Might not the Aristotelian division of the universe into two realms, the Earth and the heavens, be a fable?

Behind all these questions there loomed the ultimate problem. If the Earth is not the still center of a finite universe, what power keeps the vast mass of the Earth and all the other planets in motion around the Sun? Copernicus's theory was still couched in geometric terms, it was abstract and conceptual, but it implied the need for a new kind of physics. The Christian church had adopted the Aristotelian view of the universe and had built it into orthodox religious thought. When that view was challenged, as it now was, the challenge might appear to be directed against the entire structure of Christian thought. The implications of Copernicanism took many years to sink in and be fully understood, but this shift in humans' vision of the universe had a profound effect not only on science but on religion and philosophy.

THE SCIENTIFIC REVOLUTION

The Revolution in Astronomy: Tycho Brahe

If there is a single date that marks the beginning of the Scientific Revolution, it is November 11, 1572. Shortly after sunset on that day the Danish astronomer Tycho Brahe glanced up at the constellation Cassiopeia and was astonished to see a star brighter than any of those surrounding it, a star that, he realized, had not been there before. Hurrying home, he carefully noted its position, a procedure that he repeated on many succeeding nights, until it grew fainter and finally vanished from sight in March 1574.

This star had no tail and was therefore no comet; it did not move as a planet does; no change at all was visible in its position during the seventeen months of its appearance. Tycho concluded that it was simply a star, a new star, that had flared into intense life and then faded back into darkness. The implications of this event for traditional science were profound, for every scholar in Europe accepted Aristotle's doctrine that the heavens were eternal and unchanging, and that change occurred only on Earth. This idea had been reinforced by the religious idea that heaven was the dwelling place of God and his angels. Tycho now realized that this fundamental teaching was wrong.

New Stars in the Heavens

Tycho described the new star (which we now know was a supernova) in a short treatise, *De Nova Stella* ("The New Star") of 1573. The question it raised concerning traditional science became even stronger just four years later with the appearance of a great comet in November 1577. Tycho observed its path every night for two months and concluded that it was not below the Moon, as had always been believed, but above it, and that its path crossed those of the planets Mercury and Venus. As Tycho pondered this, he came to realize that the ancient belief in the reality of the celestial spheres must be a myth, because the comet would smash through the crystalline spheres. So in a few short years during the 1570s Tycho was driven to reject two fundamental

doctrines of medieval astronomy, and he went on to start the process of rebuilding the understanding of the cosmic structure.

Detailed Observations

Tycho Brahe (1546–1601) was a Danish nobleman who used his wealth to advance scientific knowledge. With the help of King Frederick II of Denmark Tycho created the first observatory ever built in Europe, on the island of Ven, in the strait between Denmark and Sweden. He called his observatory Uraniborg, which means "City of the Heavens."

Tycho lived just before the telescope was invented, and he used only naked-eye sighting tools; but he had instruments specially built that were larger and more accurate than any others of the time. He employed a large team of assistants and spent more than 20 years in dedicated observational work. He recorded the paths of the planets, and he remeasured the positions of many stars in order to publish a new star catalog on which accurate new star maps and globes could be based.

Tycho was aware of the Copernican theory, and he agreed that the movements of the planets could best be explained by supposing they revolved around the Sun. But he could never bring himself to accept the Copernican idea of a moving Earth, so he devised his own compromise theory of the solar system. Tycho agreed that the planets revolve around the Sun, but he argued that the Sun (and the Moon) revolved around the Earth, and that the Earth was still the center of the universe. In this scheme the paths of the Sun and Moon cross those of the planets, showing once again that the celestial spheres could have no physical reality. Tycho dedicated himself to careful observation and measurement, and he was able to lay the foundations of a new approach to astronomy.

Kepler and the Harmony of the Universe

When Tycho showed that the celestial spheres could have no real existence, he posed an immense problem for the scientific community. If the stars and planets were not carried through space on crystalline spheres, what held them in their places? Why didn't they fall to earth or whirl away into space? What power was strong enough to hold the stars in their courses? And if the Copernican theory was true, what power was strong enough to move the Earth?

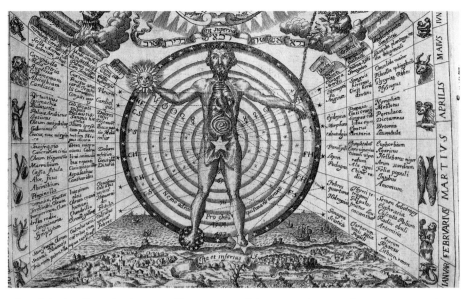

The astronomers of the seventeenth century who set out to answer these questions were all religious men, certainly not atheists, and yet none of them was content to answer simply that God was the unknown power. While accepting that God was the architect of the cosmos, they were all quite convinced that some natural, scientific explanation must be found and must be expressed in mathematical language.

This belief was especially strong in the German astronomer Johannes Kepler (1571–1630), who began the search for a new physics that would culminate in the work of Isaac Newton. Kepler's mind had a deeply mystical element. He was always searching for the underlying harmony he believed ruled the universe, and at times he believed he had found it. At the same time, he was a brilliant mathematician who spent long hours in the endless calculations that he hoped would reveal the structures of the universe.

Belief in Astrology

Throughout his life he retained a strong interest in astrology, for he believed that humans had a special place in the cosmos, that they were somehow linked to the heavenly bodies. For this reason he tried hard to find a physical basis for astrology. For example, he wondered if the quality of the light from the heavenly bodies might explain their different influences, or if the angles the planets made with each other was the crucial factor, combining their light and influence in a mathematical way.

When he was still a young man, Kepler believed that he had found the secret structure of the solar system. He discovered that the orbits of the

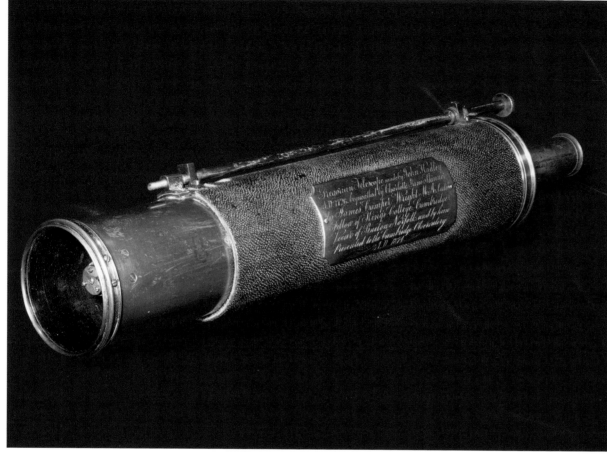

of mathematics. No one could deny the proofs that Newton had provided of the universality of gravity. But at the same time, there was a mystery at the heart of his work, for nowhere did Newton explain, or even hint at, what gravity really was and how it could operate across empty space.

For many of his contemporaries, especially in continental Europe, this was a profound difficulty. It seemed to them that Newton had had revived the idea of occult powers of sympathy, which magicians and alchemists had believed in, but which the age of empirical science had supposedly banished. They could not accept the principle of "action at a distance" which gravity implied. To this day, no particle or ray or wave of gravity has ever been discovered, although its laws are irrefutable.

Newton's work had immense philosophical implications, giving rise to the image of the universe as a miraculous mechanism, a vast clockwork machine that had been set in motion by its creator and that now functioned by immutable and perfect laws: God was to be seen in the faultless and eternal working of his creation.

THE EIGHTEENTH CENTURY
The Vindication of Newtonianism
At the time of Newton's death in 1727 the prestige of his science was immense throughout most of Europe, but there were still doubters and opponents, especially in France, where Descartes's mechanistic theories were preferred. In the course of the eighteenth century a great deal of theoretical and practical work was devoted to the business of proving or disproving certain features of Newton's theory of gravity.

Opposing Views
Perhaps the most impressive practical demonstration concerned the shape of the Earth. Newton had predicted that the Earth could not be a perfect sphere, but must be slightly flattened at the poles and slightly bulging at the equator. This was because the Earth's rotational speed was highest around the equator, and centrifugal force strongest. Descartes and his followers argued, on the contrary, that the Earth must be elongated toward the poles because it was squeezed by the vortices that they believed swept the Earth around in its orbit.

It was also Newton's view that gravity would be slightly weaker at the equator because it was farthest from the center of the Earth. This point was proved in the 1670s by several observers, when it was noticed that a pendulum clock, which kept good time in Europe, lost several minutes each day around the equator because the pendulum's acceleration due to gravity was weaker.

Measuring the Earth
But could the overall shape of the Earth be established? The French academy of science organized two expeditions to measure two arcs of a meridianal degree, one in Lapland on the Arctic circle and the other on the equator in Peru. The Lapland party was led by Pierre Louis de Maupertuis, and the Peru expedition by Charles de la Condamine, both of them leading mathematical scientists. They left Paris in 1735, but the difficulties of travel with large surveying instruments were so great that the final reports from Peru were not announced in Paris until 1744.

By meticulous surveys from mountain tops both groups measured an exact arc of a meridian line running north–south and constructed its line of curvature by astronomical observations. When this arc was extended into a circle and divided by 360, the length of a single degree was arrived at. The Lapland figure was 69.04 miles, while that in Peru was only 68.32 miles. Thus the Earth's shape must be a sphere slightly flattened at the poles, as Newton's theory of gravity had predicted.

Variations on a Theme
Another, and more profound, problem thrown up by Newton's theory concerned the dynamics of the heavenly bodies. In Newton's science the solar system was a vast, complex mechanism moving in obedience to unchanging laws. And yet astronomers had observed certain anomalies in those movements. Over a period of years the paths and speeds of the planets showed minute variations from those predicted by Newton's theory. The orbit of Jupiter, for example, appeared to be steadily shrinking, while

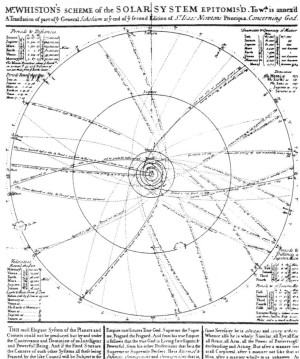

LEFT: Tiny celestial globe (with a three-inch diameter) on a brass mechanical stand for demonstrating the motion of the moon. It dates to the mid-eighteenth century.

FAR LEFT: The telescope and the microscope completely transformed peoples' picture of the complexities of the natural world. This shows the first Gregorian refracting telescope built by John Hadley in 1728.

ABOVE: A plan of the solar system that includes the highly eccentric paths of comets; it was Halley who proved that the comets were an integral part of the solar system.

LEFT: A small tabletop model of the solar system, including the moons of Jupiter and Saturn; the scale was impossible to show, of course, but the best of these models were geared so that the planets' relative periods of rotation were correct.

RIGHT: Newtonian reflecting telescope on an altazimuth stand. The telescope was made by the famous astronomer William Herschel for his friend Sir William Watson.

These complex movements were all set out by Laplace in precise mathematical terms. The solar system was a highly stable system, and it was not necessary for God to intervene to balance it. Newton's theory had been vindicated in a way that was more precise than even he could have envisaged. Between 1740 and 1790 these mathematical advances served to establish the Newtonian view of the universe beyond doubt. The rival Cartesian theory of the vortices was gradually forgotten, for the vortices could never be analyzed in this precise mathematical way; indeed, their very existence was purely hypothetical.

Scientific Laws
Pierre Simon Laplace (1749–1827) was the most important mathematical astronomer since Newton. He was the son a small farmer in Normandy, whose genius for calculation gained him a place in a military academy, and he began publishing mathematical papers before he was 20. His great work, *Treatise on Celestial Mechanics*, appeared in five volumes between 1798 and 1827, and it is one of the most technical, detailed, and forbidding works of science ever published.

Laplace was a leading figure in the French Enlightenment, whose work found special favor with Emperor Napoleon. Laplace believed that "All the effects of nature are only mathematical results of a small number of immutable laws," and his form of science clearly led toward atheism. In an interview with Napoleon the emperor pointed out to Laplace that in all his studies of the physical universe, he had never once mentioned the creator of it. Laplace famously replied, "I had no need of that hypothesis."

Laplace was also the author of the nebular theory of the origin of the solar system—that the Sun had condensed from a rotating mass of gaseous particles that then threw off the planets and their satellites. In one form or another this theory has prevailed to this day. Laplace's dying words were,

"What we know is very slight; what we do not know is immense."

Saturn's orbit seemed to be expanding. Most puzzling of all, the velocity of the Moon appeared to be steadily slowing by a fractional amount, making a month longer.

Newton himself had been aware of some of these anomalies, and he had suggested that although the universe was generally stable, it was still necessary for God to intervene from time to time to maintain its perfect equilibrium. This argument was welcomed by religious thinkers because it showed that the universe was not a self-regulating mechanism that could function without God. A succession of great mathematicians, most of them French, set themselves to analyze these problems, and some of them concluded that the ether, the mysterious invisible substance that they believed filled the universe, created a dragging effect that disturbed the movements of the planets.

Balancing Out
The true solution, however, was worked out by Pierre Simon Laplace (1749–1827), who was able to show that the anomalies in the planetary movements were cyclical and periodic, and that they balanced each other out. For example, a slight decrease in the Moon's velocity was explicable as a result of a slight increase in eccentricity of Earth's orbit, and this effect was reversible. Laplace devised a series of equations to show that the sum of all the eccentricities of the planets is invariable, and that it is distributed in proportion to the planets' masses. In effect, there is a fund of eccentricity in the solar system that is constant. If the eccentricity of one body increases, that of another will diminish, and these variations are all cyclical. The Jupiter–Saturn effect, for example, occupies a cycle of around 900 years.

Astronomy: Looking beyond the Solar System
Classical astronomy had always concentrated on the solar system, analyzing the paths of the Sun, Moon, and planets. The stars had been catalogued and marked on the chart of the heavens; but they had always been seen as mere points of light, and it had been assumed that humans could know nothing about them. Traditionally, people thought they were set in a single "starry sphere" all at the same distance from the Earth. This realm of the universe was so remote and unknowable that it was regarded in religious terms as the abode of God and his angels.

By the end of the seventeenth century the Copernican revolution and the advent of powerful telescopes had altered these ideas. It was now accepted that the Sun was a star, and the other stars were suns. This insight raised profound questions about the scale and structure of the universe. The eighteenth century saw the beginning of stellar astronomy, although at first of a rather speculative kind.

Immeasurable Universe

The first aim was to arrive at a more accurate idea of the scale of the universe. Many astronomers, from the time of Newton and Huygens, had suggested ways in which improved figures might be obtained for the distances between the Earth and the Sun and the Earth and the stars. These methods usually involved calculations based on parallax and on relative brightness. None of these figures ever came close to accuracy, but they all exploded the old view of the size of the universe that had prevailed since Ptolemy, and scientists became keenly aware that the universe was immeasurably vast.

It was Edmond Halley, who made important contributions to so many fields of science, who first questioned the doctrine that the stars were fixed and immovable. In 1718 he published observations that suggested that the stars had moved over time, making the positions given in classical star catalogs from Ptolemy to Hevelius slightly inaccurate. No general pattern in these movements was detectable, and Halley concluded that the stars were moving freely, perhaps randomly, in space.

The implication of this observation and discovery was that the stars were scattered throughout space at varying distances from the Earth, so the old idea of the celestial sphere as a shell bearing all the stars was a myth. The question then arose whether the universe had some large-scale structure, and, if so, how observation by Earthbound astronomers could uncover it and begin to understand the cosmos and ultimately prove its existence scientifically.

Changing Position

That the Sun was not in any sense the center of the universe was now accepted by all scientists, and it seemed logical to believe that the Sun itself was also moving through space, carrying the Earth and the entire solar system with it. All astronomers of this time made one important assumption—that all the stars were of approximately the same size and luminosity, and that the brightness or faintness of a star indicated fairly directly how far from the Earth it lay. For the next 50 years astronomers struggled with the near-impossible task of measuring changes in stellar positions. Before the work of William Herschel there were two major breakthroughs.

In 1729 Englishman James Bradley (1693–1762), who succeeded Halley as Astronomer Royal at the Greenwich Observatory, announced that the positions of all stars varied by infinitesimal amounts from their predicted values on an annual cycle. Some values were in advance, some behind, but in both cases they corrected themselves before creeping in again. Bradley argued that the only logical explanation was that the velocity of the Earth in its orbit around the Sun was being added to or subtracted from the finite velocity of the light incoming from the stars.

This was the "aberration of light," and it was important for two reasons.

Philos Trans MDCCX

Fig. 1.

WILLIAM HERSCHEL (1738–1822)

- Astronomer.
- Born Hanover, Germany.
- Trained as a musician and joined the Hanoverian Guards band as an oboist.
- 1755 Moved to England to work as a musician.
- 1766 Settled in Bath and started an interest in astronomy.
- Taught himself to cast mirrors and built his own reflective telescope.
- 1781 Discovered the planet Uranus which he named *Georgium Sidus* in honor of King George III.
- 1782 Appointed Astronomer Royal.
- 1787 Working with his sister Caroline, he built larger telescopes and discovered two satellites of Uranus and two satellites of Saturn, 1789.
- During his studies of the stellar universe he drew up the first catalog of double stars in 1782; proved that they orbit around each other, 1802.
- 1783 Recorded the Sun's motion through space.
- 1784 Published a paper, "On the Construction of the Heavens," revealing the Milky Way as an irregular collection of stars.
- Began a systematic search of the skies for nebulas and star clusters— discovered 2,500—published the data in three catalogs, 1786, 1789, and 1802.
- Made distinctions between the different types of nebulas.

LEFT: Herschel cast his own 48-inch-diameter mirror for this giant 40-foot-long reflecting telescope.

RIGHT: The "Alarming Comet of 1835" published in 1839, a lithograph by A. Ducote. It shows the comet—Halley's comet—with the face of Sir Edmund Halley, who calculated the comet's orbit. He had determined that the comets of 1531 and 1607 were the same object with a 76-year orbit. Halley didn't live to see his prediction come true, dying in 1742, 16 years before the next sighting of his comet.

First, it gave the first physical proof of the Copernican theory that the Earth is moving; second, given the immense velocity of light, it indicated just how vast the distances to the stars must be.

Traveling Through Space

Another English astronomer, John Michell (1724–93), drew attention to double stars—binaries as they are now called—and argued from them that gravity operated far outside the solar system. In 1760 the German scientist Tobias Mayer (1723–1762) announced his discovery of a very important principle. If the Sun, and the Earth with it, were moving through space, then we should be able to detect this as stars appearing to open before us and close behind us. This effect is exactly what happens if we walk through a wood: Trees in front that had appeared close together seem to widen out before us; but, if we look behind, the wood has assumed its dense appearance again.

Mayer argued that there should exist a "solar apex" in the direction of travel and a "solar antapex" behind. This was soon confirmed, and it was established that the Sun and the solar system are moving in the direction of the constellation Hercules. The discoveries of Bradley and Mayer were a considerable triumph of astronomical thought, overturning the doctrine of the fixed stars that had been universally believed just a few decades before.

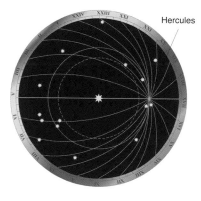

Hercules

LEFT: Herschel's drawing of the motion of the Sun and solar system toward the constellation Hercules, as suggested by Tobias Mayer.

Divine Influence

This new interest in stellar astronomy stimulated many theories about the size and structure of the universe. Among them was the highly imaginative scheme of Thomas Wright (1711–86), an English navigator and surveyor who combined scientific vision with religious belief. Wright took the Newtonian principle of gravity and the newly discovered movement of the stars, and wove them together into a new cosmic structure.

He proposed that the universe contained a "divine center" that acted as a gravitational center around which all the stars, including our own Sun, move in orbit. This explained why the universe did not collapse into a single body under the force of gravity, a problem that had worried many post-Newtonian thinkers. The stars, he thought, were set into a series of concentric shells, all revolving around the divine center, thus accounting for the stellar motions seen by Halley and for the aberration of light. One might see Wright's vision as a huge stellar equivalent to the solar system, magnified on a vast scale, but still working under the influence of gravity, and still with a clear center and a spherical structure.

Wright made no contribution to precise astronomy in the way that Bradley or Mayer did, but his book *An Original Theory or New Hypothesis of the Universe*, published in 1750, influenced many other thinkers. It shows that cosmological thought was moving in new directions. Wright was trying to come to terms with a universe whose scale was vast, and possibly even infinite. But he felt still that it must possess a clear and rational structure and be under divine control. Wright, like William Derham, believed that God's power now exerted itself over a far vaster universe than just this solar system, and that many worlds existed scattered throughout space.

New Planets and New Stars: William Herschel

The universe of the eighteenth-century astronomer did not consist only of stars and planets. Comet hunting became a compulsive activity because these erratic, wandering bodies were now understood to be part of the solar system and to conform to Newton's orbital laws.

It was a dedicated French comet hunter, Charles Messier (1730– 1817), who turned his attention to another group of objects in the night sky. They were certain unusual stars that ancient astronomers had described as "cloudy" or "milky"—the nebulas. Messier wanted to chart them in order to avoid any possible confusion with comets, and by 1784 he had catalogued 101 of them. They are still known according to their M numbers: The one in Orion is M42, in Andromeda M31, and so on. Messier had no idea what the nebulas were, and his interest in them was limited (although he produced some intriguing first drawings of their shapes). But his studies provided the starting point for a fresh field of astronomy, one that, much later, in the twentieth century, would become central to our understanding of the universe.

Herschel's Contribution

Messier's work on nebulas was enormously extended by the greatest observational astronomer of the eighteenth century, William Herschel (1738–1822). Born in Germany and originally a musician, he moved to England when he was only 19. Herschel trained himself in both astronomy and in making his own telescopes, to such effect that by 1780 his instruments were the best in Europe.

In March 1781, using a seven-foot reflecting telescope, Herschel identified a body in the sky that he knew was not a star, although it might have been a faint comet. Over the following months, however, observations convinced

Herschel that it was a planet, the first new planet to be added to the solar system since the dawn of astronomy 5,000 years before. Patriotically, Herschel wanted to name it Georgium Sidus ("George's Star") after the king of England, but other astronomers found this completely out of keeping with the classical names of the other planets. Some referred to it as "Herschelium," but eventually, the name Uranus was agreed on, in Greek mythology the father of Saturn and grandfather of Jupiter. Herschel became famous, was awarded a royal pension, and was able to build even larger telescopes and turn his attention his long-term goal in astronomy.

Ambitious Undertaking

Herschel's goal was nothing less than an attempt to map the universe in three dimensions. All conventional star charts to that date had treated the heavens as a two-dimensional plane in which objects are separated only by the angular distances between them. Herschel wanted to survey the radial distances of the stars from the Earth and their distribution, and so offer a picture of the universe in depth to determine whether there was a large-scale structure such as that envisaged by Thomas Wright.

Herschel had at that time no reliable data about stellar distances, but he made the basic assumption that all stars are more or less equal in brightness, and that the luminosity that we perceive is therefore a result of their distance from the Earth. Herschel then proceeded to count stars—thousands and thousands of stars—dividing the night sky into small set areas that he called "gages." He continued with this work throughout the 1780s, and he reached the conclusion that the stars were grouped together in a vast and rather irregular disk formation. The concentration of stars that we see as the Milky Way is our edge-on view of the disk, while when we turn away from the disk's center, the stars noticeably thin out. In this he was correct, although he supposed that all the objects that he saw in the heavens formed part of a single system.

Deeper into Space

In 1789 Herschel began using the largest telescope of his own construction, a 40-foot reflector. It was so powerful that he began to have doubts about his model of the

universe, for wherever he turned his new instrument, more stars were brought into view. He began to see that his earlier survey had not penetrated to the limits of the universe, as he had hoped. What he did not conceive was that some of the objects that he saw might be part of further and more distant systems. Among these objects were the nebulas, to which he paid special attention. He could distinguish that some nebulas were composed of individual stars, while others were truly nebulous or cloudlike. Because they appeared in various intermediate stages between stars and clouds, Herschel suggested that what was revealed here was the life cycle of the stars. As in the nebular theory advanced by Laplace, the clouds of gas were thought to be in the process of condensing into stars.

In this Herschel was mistaken, for most of the nebulous masses that he was seeing were indeed star groups too distant to be resolved. They were, in fact, galaxies beyond and separate from our own. Nevertheless, Herschel had proposed the idea that change on a cosmic scale was at work in the farthest corners of the universe. This was a profound shift from the static, mechanistic view of the universe that had prevailed at the beginning of the eighteenth century. In some ways it can be seen as the counterpart to the new Buffon-style view of nature here on Earth as in a process of gradual change. Despite the limitations of his time, William Herschel's work stands at the beginning of modern observational cosmology.

URBAIN LEVERRIER (1811–77)

- Astronomer.
- Born in St. Lô, France.
- 1836 Became a teacher of astronomy at the Polytechnique.
- Interested in the motions of the planets, he published *Tables de Mercure* ("Tables of Mercury") about the planet Mercury.
- 1846 Admitted to the French Academy.
- Deduced the existence of an unseen planet by observing irregularities in the motion of the other planets, he calculated its exact position; a few days later, exactly where he predicted, Johann Galle discovered the planet Neptune, 1846.
- 1849 Elected to the French Legislative Assembly.
- 1852 Napoleon III made him a senator.
- 1854–70 Worked as director of the Paris Observatory.

THE NINETEENTH CENTURY

The Changing Solar System

The main achievement of nineteenth-century astronomy was in its investigation of the universe of the stars and in revising people's ideas about the scale of the universe. But the more traditional study of the solar system could still produce its own revolutionary discoveries.

In 1772 two German astronomers, Johann Titius (1729–96) and Johann Bode (1747–1826), had drawn attention to a curious sequence in the planets' distances from the Sun. The numerical sequence runs:

$$0 + 4 = 4, 3 + 4 = 7, 6 + 4 = 10, 12 + 4 = 16,$$
$$24 + 4 = 28, 48 + 4 = 52, 96 + 4 = 100, 192 + 4 = 196.$$

If the totals in each of these sums are divided by 10, the result is extremely close to the distance of each planet from the Sun measured in astronomical units, that is, the Earth's distance from the Sun:

Mercury: 0.4, Venus: 0.7, Earth: 1.0, Mars 1.6,
Jupiter: 5.2, Saturn: 9.5, Uranus: 19.2.

The fit is generally very good, but there was evidently a gap between Mars and Jupiter, and it was naturally asked whether an undiscovered planet could possibly be orbiting the Sun at 2.8 astronomical units distance. For some years astronomers studied this possibility, and in 1801 the Italian observer Giuseppe Piazzi (1746–1826) did indeed locate a planetary body in exactly the predicted orbit; he named it Ceres. Yet it proved to be very small, having a diameter of only 600 miles—roughly the size of Spain.

The situation became still more puzzling when, in the years that followed, other even smaller bodies were found to be sharing the same orbit. By 1872 more than 100 "asteroids" had been located, and it was long believed that they must form the remains of a large planet that had somehow disintegrated. The status of the Bode-Titius law is still unclear. Neptune does not fit the pattern well, and with Pluto it breaks down completely. It appears to be a purely numerical relationship and not a physical law at all, but it had played a part in the discovery of a previously unknown feature of the solar system.

New Planet

In the 1820s astronomers also began to be puzzled by the orbit of Uranus, the outermost of the known planets, discovered by William Herschel in 1781. Its orbit displayed small irregularities that suggested either that Newton's laws of gravitation did not apply rigidly at such a great distance from the Sun, or that the orbit was being disturbed by another body. Could there be another planet beyond Uranus? Using the mathematical models set out by Laplace, astronomers set to work to predict the likely position of such a body and then to scan the night sky in the hopes of finding it. This led to one of the most contested "first" disputes in the history of astronomy. In September 1846 the French astronomer Urbain Leverrier (1811–77) submitted detailed predictions of the unknown planet's path to the Berlin Observatory to be used as a basis for searching, and within days the new planet was found.

But the international celebrations that followed became confused when it was announced that a young English mathematician, John Couth Adams (1819–92), had reached exactly the same result a whole year earlier. Adams had sent his calculations to the professor of astronomy at Cambridge University and to the astronomer royal, but they had not followed them up with an optical search of the sky. National rivalries between English and French scientists played a part in this story, but today both men are credited with the discovery of the new planet, which was named Neptune, god of the deeps of the sea, because it lay so deep in space.

Written in the Stars

There were two curious sequels to this discovery. First, Leverrier turned his attention to the orbit of Mercury, which had also shown certain irregularities that had puzzled Newton and many other scientists. Leverrier predicted that another planet must lie inside Mercury's orbit, even closer to the Sun, and he decided to name it Vulcan after the god of fire; it was never found.

The second consequence was of little direct importance to science, but it affected the astrological community. Traditional astrological ideas about the influence of the planets had always been concentrated on the five planets known since ancient times. The discovery of Uranus and then of Neptune at first raised huge problems for astrologers: Surely these planets must also influence human affairs, and surely all previous astrology was made invalid because its practitioners had not even known the true number of planets.

Since the late nineteenth century astrologers have set about describing the characteristic influences of the new planets (and of Pluto), but those characteristics relate to the role of the gods Uranus, Neptune, and Pluto in classical mythology. However, we know that the naming of these new planets was really an arbitrary process: They might easily have been given different names and therefore would presumably have been given different qualities by the astrological community. This problem of the new planets was to prove a serious one for modern astrologers.

Probing the Universe of Stars

William Herschel had laid the foundations of a new approach to stellar astronomy, and the leading astronomers of the nineteenth century devoted themselves to the task he had started—redrawing people's intellectual map of the universe. Herschel's son John (1792–1871) continued his father's work to the extent of observing and cataloging 2,300 nebulas by the year 1833. In the process he made a striking and important drawing of the nebula M51, which lies in the constellation Canes Venatici. He saw it as a central cluster surrounded by a divided ring of stars. He realized that this ring, if seen by

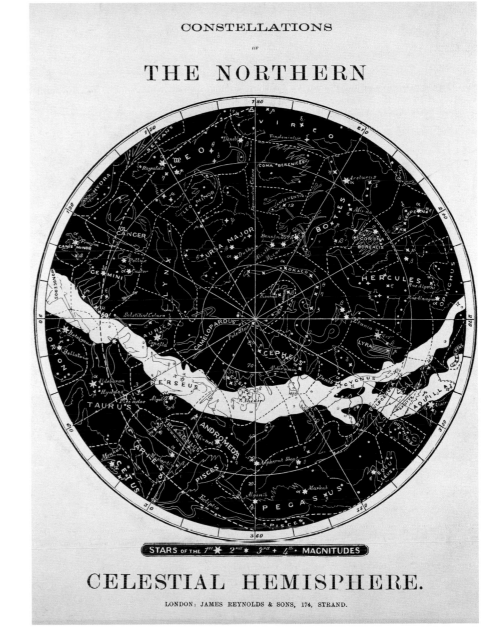

LEFT: Rosse's giant reflecting telescope. The mirror was 72 inches across, and Rosse made important observations with it, despite the cloudy and unfavorable climate of central Ireland where it was located.

BELOW: A spectrum of sunlight: Different stars produce different patterns of dark lines ("absorption lines") depending on the chemicals present; these patterns act like a fingerprint showing the constitution of the stars.

more than a year before concluding that it showed a parallax movement of one-third of one second of arc. This distance then became the base of a very thin triangle, two of whose angles were known. When that base was stated as the diameter of the Earth's orbit, it became possible to derive a tentative figure for the star's distance from the Earth; Bessel gave it as just over 60,000,000,000,000 miles. In modern terminology this equates to around 12 light-years and is a very good estimate.

Within a year or two other stellar parallaxes were found. A Scottish astronomer, Thomas Henderson (1798–1844), then working at the Cape Observatory in South Africa, announced that the brightest star in the constellation Centaur had a parallax more than double that of Bessel's chosen star, meaning that its distance from the Earth was less than half, as indeed it is: The star Alpha Centauri, at four and a half light-years, is generally agreed to be the star closest to Earth. The magnitudes of these figures were beyond anything that scientists had anticipated, and they were for stars that were relatively close to the Earth. There seemed to be no answers in sight to the problem of comprehending the scale of the universe.

Unlocking the Chemistry of the Universe

In 1844 the French philosopher Auguste Comte (1798–1857) was discussing the limitations of scientific knowledge, and he gave his opinion that man would never know anything about the stars except what they looked like from a distance; we could never know about their physical or chemical natures. Within fewer than 20 years of this prediction Comte was to be proved wrong by one of the most important, and perhaps unplanned and unexpected, technical breakthroughs in astronomy—the technique of spectroscopy.

The fact that light from flames can be analyzed by a prism and will show different colors depending on the substance being burned was realized by a number of scientists early in the nineteenth century. In 1814 a Munich instrument maker, Joseph Fraunhofer, noticed that in these cases the spectrum was intersected by a number of dark lines whose meaning he was unable to explain, although he saw that the pattern of the lines varied as different substances were burned.

Color Theory

The chemical significance of the spectrum was grasped by, among others, two Englishmen, the physicist and photographic pioneer William Fox Talbot (1800–77) and astronomer John Herschel (1792–1871). Talbot wrote that "a

an observer at the center, would be similar to the Milky Way as seen from the Earth. "Perhaps," he wrote, "this is our brother system." The older Herschel had seen only the stars visible from England; but in order to complete his father's work, John transported his great 20-foot reflecting telescope in 1833 to a site near Cape Town in South Africa, where he spent the next four years surveying the southern skies. On his return to England he published a new catalog of 1,700 more nebulas, making it appear, as one commentator remarked, that the study of the nebulous heavens "was the exclusive domain of the Herschel family."

Seeing Stars

This private monopoly was broken in the early 1840s when William Parsons, later earl of Rosse, built a gigantic reflecting telescope at his home in Birr, Ireland, with a mirror six feet in diameter, easily the most powerful instrument of its time. Within weeks of its first use Rosse had produced a drawing of the M51 nebula superior to that by Herschel and showing its

characteristic spiral form. The great question about these nebulas was whether they were really composed of stars, or whether they were simply clouds of gas. Both Herschel and Rosse had clearly resolved M51 into a mass of stars, and its distinctive shape suggested strongly that it was a star system. But was it a subsystem within our own or separate from our own? There was still no way of answering these questions because no method of measuring or even estimating stellar distances had been discovered.

A significant step in that direction occurred when in 1838 the German astronomer Friedrich Wilhelm Bessel (1784–1864) succeeded in doing what had eluded all astronomers to that date, namely, measuring the parallax movement of a star caused by the Earth's motion in its orbit around the Sun. With the best instruments available Bessel carefully observed a star in the constellation Cygnus for

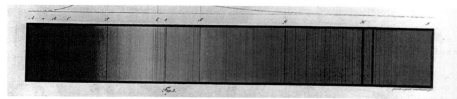

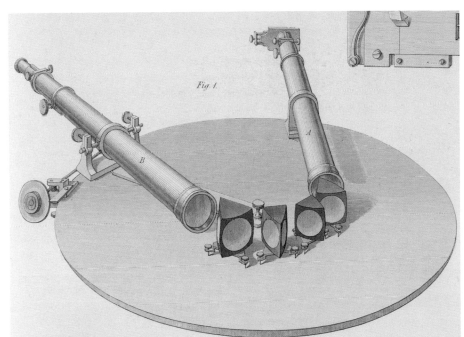

Fig. 1.

HEINRICH OLBERS (1758–1840)

- Astronomer.
- Born in Ardbergen, Germany.
- Studied medicine at Göttingen and Vienna.
- 1779 Worked out a way of calculating the orbits of comets.
- 1781 Set up in medical practice in Bremen, but studied astronomy at every opportunity from his small home observatory.
- 1802 Discovered the minor planet Pallas and another, Vesta, in 1807.
- 1815 Discovered a comet that returns every 70 years that was named for him.
- 1826 The Olbers' paradox—why is the sky dark at night when the stars and galaxies are so bright? His answer to his own question is that the universe cannot be an infinite static arrangement of stars; this led to the theory of the expanding universe.

LEFT AND BELOW LEFT: Two views of a spectroscope: Light from the subject is refracted through one or more prisms and studied through the telescopic eyepiece.

---

presence of sodium, calcium, magnesium, iron, chromium, and copper. The Sun's chemical composition could be analyzed after all, and it turned out to be made of the same elements of which all earthly matter was built.

Spectral Analysis

A further startling conclusion was that the Sun (and therefore by extension the other stars) had an atmosphere. This discovery arose from the dark lines ("absorption lines") that crossed the colors of the spectra. Kirchhoff realized that these lines must be caused by bands of light of a given wavelength being absorbed by the same elements that existed as gases in the intense heat surrounding the Sun.

When the light from burning sodium or magnesium, for example, encountered light from those elements burning as gases, the two canceled each other and produced the characteristic dark "absorption lines." It was the pattern of these lines combined with the pattern of colors in the spectrum that functioned like the fingerprint of any given star. The existence of the Sun's atmosphere came to be appreciated when the first photographs of the Sun revealed its huge glowing corona, which is its atmosphere, extending for millions of miles into space.

Kirchhoff wrote a series of papers in which he explained their discovery and rationalized what was being seen in the different spectra. A monument to Kirchhoff that stands in Heidelberg commemorates the discovery of spectral analysis that "unlocked the chemistry of the universe." Within a few years astronomer Norman Lockyer (1836–1920), in England had used the spectrum to identify an unknown element in the Sun, which he named helium, and which was not isolated on Earth until 1895. Spectroscopy was to yield another momentous discovery—that it could be used to measure the velocity with which the stars were moving.

Doppler Effect

This was an application of the principle described by the Austrian physicist Christian Doppler (1803–53) in 1842 that a moving source of energy produces radiation of a longer wavelength if its receding and of a shorter wavelength if it is approaching. That is because the waves of radiation do not arrange themselves symmetrically around a moving energy source: They are "squeezed" in front of it and "stretched out" behind it. When applied to the light from a star that is receding from the Earth, this means that the long-wavelength part, the red end of the spectrum, is stretched, while the short wavelength blue end is contracted. This principle has become known to science as the "red shift." If a star were rushing toward the Earth at great speed, the reverse effect would occur—there would be a blue shift.

In 1868 the English astronomer William Huggins (1824–1910) analyzed the light of Sirius, the brightest star in the sky, and found that its red was shifted to the extent that he was able to calculate that the star was moving away from the Earth at a speed of almost 30 miles per second. This

glance at the prismatic spectrum of a flame may show it to contain substances which it would otherwise require a laborious chemical analysis to detect."

In Germany Robert Bunsen (1811–99) developed his famous gas burner to burn with a colorless flame so that it would not interfere with this color analysis. It was his colleague Gustav Kirchhoff (1824–87) who suggested that he should refract the light through a prism instead of looking at it through various colored filters. These two chemists, working in Heidelberg in the 1850s with their Bunsen burner and prism, realized that each element as it burned emitted a spectrum of bright colors intersected by dark lines as individual as a fingerprint, and they discovered that this technique could be transferred from combustion in a laboratory to the realm of the stars.

The story is told that this discovery followed their observation of a distant fire that they saw from their laboratory window. With their spectroscope they were able to tell that among the substances that were burning were the metals barium and strontium. Some time later Bunsen suggested to Kirchhoff that if they could analyze the material in a distant fire, why should they not do the same for the Sun? When they turned their spectroscope on the Sun, they were astonished to find a spectral pattern exactly similar to what they received from any laboratory combustion. In a few months in late 1859 and early 1860 they had mapped the spectrum of the Sun's light to such an extent that they had detected the

LEFT: Giant reflector telescope in Dublin in 1881.

ABOVE RIGHT AND FAR RIGHT: Two striking early photographs of the Moon, taken by Warren de la Rue on a 13-inch reflecting telescope in the middle of the century (around 1858) at an observatory just outside London.

technique of analyzing the red shift was to assume the utmost importance in the twentieth century, when it was applied to galaxies outside our own. Spectroscopy became one of the basic tools of the astronomer when examining any star, and its findings helped build an entirely new and unexpected view of the cosmos.

Traveling Light

Nineteenth century astronomy can be seen as an unfolding series of clues or stepping stones toward a new view of the cosmos, which finally emerged in the 1920s and 1930s. Another was the so-called "Olbers Paradox," named for the German astronomer Heinrich Olbers (1758– 1840), although he was not the first to discuss it. Olbers asked simply why the sky is dark at night, given that the heavens are infinite, and the distribution of stars is even. The

RIGHT: Olbers's paradox. Stars are evenly scattered through space, and although the distant ones are fainter, there are more of them; therefore, the night sky should be evenly lit with starlight.

Observer

amount of light reaching the Earth from distant stars is very small, but on the other hand, the number of stars also increases with distance, so why isn't the night sky glowing with an even light?

Olbers thought there might be a kind of dust in space that absorbs the light, but the answer was divined by another German, Johann Mädler (1794–1074). The light from all the stars that were out there had simply not had time to reach the Earth. But if this were true, then the time of the light's travel must be less than the age of the universe, and given the immense speed of light, this had profound implications for ideas about both the age and the size of the universe.

New Techniques and New Horizons in Astronomy

The second half of the century saw the observational basis of astronomy transformed by two new techniques of data analysis—photography and spectroscopy—while the ever-increasing power of telescopes revealed more and more about the stars and planets.

In telescope design there were persistent difficulties with both types of instrument, the refractor and the reflector. As the lenses of refractors grew larger and more powerful, the tubes had to grow to accommodate their focal length, and it became more difficult to maintain rigidity to prevent the tube from flexing under its own weight and so distorting the image. Between 1840 and 1890 the tubes of these refractors, almost 100 feet long, with lenses three feet in diameter, reached the physical limit of their development.

The alternative type of telescope, the reflector, was much shorter, but the principal difficulty was the mirror, which had to be cast from metal in a precise concave form and polished to gather and reflect the light. They were enormously heavy and difficult to cast, and they tarnished easily. The potential for much larger reflectors developed after the 1860s, when physicist Jean Foucault (1819–68) in Paris pioneered the technique of silvering glass mirrors. They were lighter, easier to cast and easier to polish than metal mirrors, and by the end of century, the reflector was undoubtedly the telescope of the future.

New Sightings

The improved power of the telescope yielded its clearest results in planetary astronomy—which could still reveal some surprises. As late as 1895 the rings of Saturn were proved to be neither solid nor liquid, but composed of meteor-like particles, when James Keeler made the necessary observations at the University of Pittsburgh Observatory. The composition of comets was analyzed, and it became evident that their heads did not merely reflect sunlight, but emitted their own radiant light, although how they did so was not clear.

Perhaps the greatest interest centered on Mars. In 1877 it approached very close to the Earth, as it does about every 15 to 17 years. In that year Asaph Hall (1829–1907), astronomer at the U.S. Naval Observatory, was the first to observe two satellites, which he named Phobos and Deimos, in Greek "fear" and "panic"—the attendants of the war-god. They are extremely tiny,

respectively only 25 and 15 kilometers (15.5 and 9.3 miles) in diameter, and Hall explained he was able to detect them because he had calculated from Mars's mass and gravity where their orbits might lie.

There is a curious history behind the story of Mars's satellites, for their existence had been predicted in the seventeenth century by Kepler. On the basis of a numerical progression from the Earth's one satellite to Jupiter's four (then known) he felt that Mars should have two. This idea became widely known, and their discovery was predicted by Jonathan Swift (1667–1715), the Anglo-Irish clergyman, in *Gulliver's Travels*. Mars became even more a focus of attention when, soon after the 1877 approach, the Italian astronomer Giovanni Schiaparelli (1835–1910) announced his claim that he had observed artificial canals on its surface.

This claim was taken up in the 1890s by the American Percival Lowell (1855–1916), who supported the idea that intelligent beings on Mars had built the canals to irrigate the planet with melt-water from polar ice. The possibility of life on Mars continued to fascinate science writers for a century. Lowell spent many years trying to locate an unknown planet believed to lie beyond Neptune, which was finally discovered in 1930 and was named Pluto.

Preserved Image

Even the most powerful telescopes could reveal little directly about the individual stars, but the application of photography to astronomy did have a huge effect on cosmology. Beginning in the 1850s, the Moon and Sun were naturally the first celestial objects to be photographed. The Sun's corona, seen during eclipses, attracted special interest because its immense glowing halo suggested that the Sun had an atmosphere. But it was when photographs were taken through powerful telescopes of star fields that its importance was really revealed, for a time-exposed photographic plate could detect light sources far fainter than the human eye, and it could preserve the image for future study.

In 1882 a Scottish astronomer at the Cape Observatory in South Africa, David Gill (1843–1914), made some excellent photographs of a bright comet then passing through the sky. But when the plate was exposed, no less striking than the comet were the number and clarity of stars in the background. It was as a result of photographs like these that a conference in Paris in 1887 decided to create a new series of celestial maps of unprecedented detail, not drawn by hand but in photographic form. The international *Carte du Ciel* (Map of the Heavens) project took many decades to complete, but it revolutionized the practice of stellar astronomy because these plates could be studied by all scientists years after they were taken. The astronomer was no longer compelled to spend endless nights watching the sky, and special instruments were built to measure and coordinate the photographic plates. These time-exposed photographs required a clockwork mechanism to be geared to the telescope in order to hold a field of view, sometimes for an hour or more.

Photography was especially important in the study of the nebulas, in determining whether they were star fields or gas clouds. Henry Draper in America and Isaac Roberts in England made many historic photographs of nebulas, some of which are, as we now know, distant galaxies in their own right.

Colorful Universe

The technique that had the most profound implications for astronomy and cosmology was undoubtedly spectroscopy, which analyzed the light from stars. Pioneered in the late 1850s by Bunsen and Kirchhoff within a decade the spectra of several thousand stars had been recorded and compared, and it became evident that they fell into groups. The Italian astronomer Angelo Secchi (1818–78) considered that there were four basic types, which ranged from those in which the white-blue light predominated to those in which the yellow-red end of the spectrum was most evident. These types could be explained simply as showing the different temperatures of the stars. The hotter the star, the more blue-white light; the cooler, the more yellow-red.

Sharp-eyed observers had long known that the stars were of different colors —Ptolemy had spoken of "golden-red Arcturus"—and now here was a clear scientific explanation. In the 1880s and 1890s a team of observers at Harvard College Observatory under American E.C. Pickering (1846–1919) examined thousands of spectra and expanded Secchi's four types into ten. The implication began to dawn on astronomers that these spectral types might represent the stages in the life-cycle of a star. Stars were perhaps evolving, cooling from white heat to dull-red heat, just as metals from a furnace do. And just like those metals, the process was revealed in their spectra, only in the case of stars it must occur over eons of time.

The implications of this insight, and of the photographs of nebulas and star-fields, would be fully drawn out only in the twentieth century, when they provided one of the foundation stones for a startling new vision of the age of the universe. Astronomy was still an observational and deductive science, for

it could not handle and experiment with its subject matter in the way that chemistry or physics could. Yet new and unexpected links with chemistry, with photography, with new techniques in optics and engineering had enormously enlarged the data available to astronomers and had brought their science to the threshold of a new age.

THE TWENTIETH CENTURY

Twentieth-century astronomy has given birth to a set of radically new ideas about the structure of the universe. These ideas took shape in a series of steps in which new observations and new techniques of interpreting them showed scientists how they might understand the scale of the universe and even perhaps how it had come into being. The findings of chemistry and physics were increasingly used to build up models of the processes at work in stars and between stars, thus creating the new science of astrophysics.

Star Catalog

Spectroscopy—the study of the emissions of light and energy given off by substances—was the first clue. At Harvard in the 1890s a team of astronomers led by Edward Pickering (1846–1919) was engaged in a massive program of analyzing the spectra (light waves) of thousands of stars. One of the astronomers, Annie Cannon (1863–1941), noticed that these spectra fell naturally into around ten types, from those in which the blue light predominated, to those where the red was dominant, and in which the dark lines that showed the presence of the various elements fell into patterns. In 1901 a catalog of over a thousand such spectra was published, and 20 years later the Harvard star catalog had grown to number over 225,000, all confirming the same pattern.

Many astrophysicists saw at once that these spectra revealed the surface temperature of the stars, and two astronomers, working independently, set out to relate these temperature spectra to the visual magnitudes, (the brightness) of the stars. The Dane Ejnar Hertzsprung (1873–1967) and the American Henry Russell (1877–1957) both produced diagrams showing that the great majority of stars were grouped into one main sequence moving from high temperature and high luminosity (light) to much lower levels of both. Further analytical techniques enabled them to calculate the relative sizes of many stars, and this produced two important subgroups that broke the predominant pattern: One group of stars was very hot, but far less bright than they should be because they are very small, while another group was low-temperature but very bright because they are exceptionally large.

Star life cycle

What did all this mean? Did it mean that there existed a number of fundamentally different types of stars functioning in different ways? Both Hertzsprung and Russell took a different view, for they concluded that these spectra showed that stars were linked in a process of evolution: What they were seeing in each case was a star at one particular stage in a cycle through which all stars would pass. But which way was the cycle moving? Russell thought that stars began as large, red, relatively cool bodies and grew steadily hotter and denser, while Hertzsprung took the opposite and correct view that stars were gradually cooling.

The Hertzsprung-Russell diagram first appeared in 1913 and quickly became one of the astrophysicist's fundamental tools, showing the place of any star in the evolutionary tree of the cosmos. The classical belief that the heavens were eternally unchanging had long ago been abandoned, but the possibility that the stars themselves were dynamic—they were, to put it simply, living and dying—was a startling new idea.

Atomic Fusion

But what was actually happening as the stars burned hotter or cooler? The fire in the Sun and stars that could burn for thousands, or perhaps even millions, of years without being consumed had long been a mystery to astronomers. Before the advent of atomic physics no known mechanism could account for the stability of the Sun. It was the British astrophysicist Sir Arthur Eddington (1882–1944) who applied atomic theories to the stars. Eddington used Einstein's discovery of the equivalence of mass and energy (see below) to suggest that the source of solar energy is atomic fusion, that hydrogen atoms were fusing to helium, releasing huge amounts of energy. In his classic book The Internal Constitution of the Stars (1926) he also calculated the mass of the Sun and suggested a lifetime of billions of years, which was in line with the processes embodied in the Hertzsprung-Russell diagram. As Rutherford's work on the atom had opened up new ideas on the age of the Earth, Eddington applied the same principle to the Sun and stars.

Eddington was not able to put forward a precise model of the atomic transformations occurring inside a star, and it was some years later that Hans Bethe (1906–2005), one of the many refugee scientists from Germany working in America, gave a more detailed answer. Bethe showed that six separate atomic transformations were required, leading from hydrogen to helium, with carbon acting as the vital catalyst. Further detailed studies would show that the hottest stars were the youngest, and that they ended their life as the red giants and white dwarfs that formed the important subgroups in the Hertzsprung-Russell diagram. Eddington and Bethe had shown that the structure and behavior of the atom, so bizarre, unexpected, and powerful, turned out to hold the key to large-scale cosmic processes too.

To Infinity and Beyond: The Scale of the Universe

Before humankind could understand the structure of the

LEFT: The Andromeda Galaxy, M31, taken by Isaac Roberts on December 29, 1888. It was taken on a twin telescope—a 20-inch reflector and a seven-inch refractor. It was taken with an exposure of four hours from a home observatory.

BELOW: The Hertzsprung-Russell diagram, which relates star temperature to luminosity, from low levels in the bottom right to high levels at the top left.

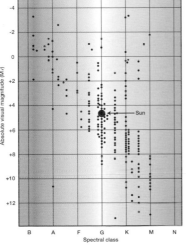

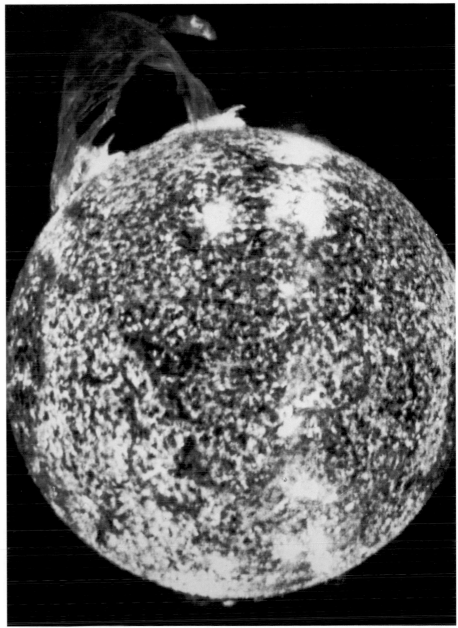

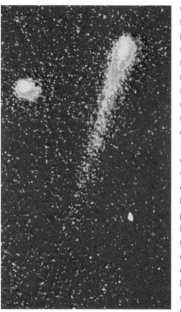

universe, it faced the fundamental problem of grasping its scale. All the thousands of objects that filled the sky, and the millions more that were being revealed by powerful new telescopes, lay scattered through space at vastly different distances from the Earth. But were all these objects part of one system, or were there distinct groupings or levels among the heavenly bodies? This was the old problem that Herschel had first addressed: How to map the heavens in three dimensions, a far more difficult task than simply locating points on the celestial sphere. For the nearer stars the established technique of parallax calculation could give a reasonable idea of stellar distances, but for the more distant stars some quite different measuring guide was needed.

Variable Stars

Such a guide was discovered in 1912 by Henrietta Leavitt (1868–1921), who was another astronomer working on the Harvard star catalog under Edward Pickering. Leavitt was studying a group of variable stars—stars whose brightness rises and falls over a period usually of several days. In the stars she was studying, which were all located in the Small Megellanic Cloud deep in the southern sky, she noticed that the longer the period over which this change took place, the brighter the star was. This relationship turned out to be quite precise: They were like preset beacons, so that just by observing its period, one could tell how bright it was. Therefore, if such a variable star were seen to have a very long period, but it appeared very faint, this difference between its apparent brightness and its true magnitude would indicate how distant it was.

This new technique was taken up by several astronomers in an attempt to rescale the distribution of stars in the heavens. The most prominent was the young American Harlow Shapley (1885–1972), who was on the staff at Mount Wilson Observatory using the 60-inch reflector, then the largest telescope in the world (the 100-inch was not completed until 1918). By studying examples of these variables at the farthest limits of visibility,

Shapley was ready in 1917 to announce his finding that the Milky Way Galaxy must have a diameter of approximately 300,000 light-years (in one light-year light travels a distance of about 6 million million miles).

In its time this was a staggering figure, ten times larger than that suggested by any other astronomer. He considered that the galaxy was shaped like a flattish disk with a bulging center, and that the Sun and our solar system were located near the edge. Shapley's conception of the scale of our galaxy was broadly correct, but it had one unfortunate result: Its size was so immense that Shapley considered that the galaxy was identical with the universe, that even the faintest objects visible in the heavens were contained within this grouping, and that he had therefore solved the fundamental problem of the size and structure of the universe.

Star Systems

Most astronomers were reluctant to accept Shapley's view because its scale was so revolutionary, but there were a few who disagreed for other reasons. Among them was Heber Curtis (1872-1942) of the Lick Observatory in California, who had been making a special study of the spiral nebulas. Curtis felt instinctively that these nebulas (vast indistinct clouds of stars) were self-contained star systems outside our own galaxy, but the difficulty was to find some distance marker to settle the problem.

Between 1915 and 1920 Curtis was able to identify a number of novas—exploding stars that are among the brightest objects in the heavens—in some of these nebulas. This established in the first place that the nebulas were definitely composed of stars, while a comparison of their very faint visibility with the tremendous brightness of novas that were already known showed that they must be at least a hundred times more distant than any novas ever recorded. Using the brightness of such a nova in the Andromeda nebula, Curtis estimated that it lay 500,000 light-years away, further than even than Shapley would allow, but with the crucial difference that this nebula was an "island universe" in its own right, another star system comparable to our own galaxy.

Curtis and Shapley met in a public debate in April 1920 at the National Academy of Sciences in Washington to present their differing views. This meeting has become famous in the history of astronomy and is often referred to as "The Great Debate," but in fact it was scholarly and inconclusive. Both men were only partly right—Shapley on the immensity of our galaxy, Curtis on the separate identity of the nebulas such as the Andromeda—but they had succeeded in defining the major problem that cosmologists would debate for the next decade.

This problem of the scale of the universe was developed on a technical level, using new and sophisticated techniques of observation and measurement. But more than that, it was also an intellectual quest, an attempt by humankind to remap its universe and to try to place itself within it. Just as the atomic physicists of this period were revealing the unbelievable

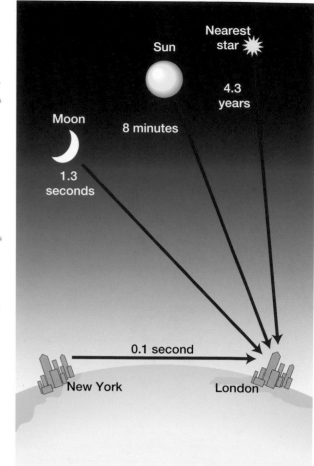

complexities of matter's material components, so cosmologists were grappling with a new vision of the unsuspected scale of the universe. It was a very exciting time to be an astronomer, and the final picture that emerged would exceed anybody's anticipations.

Parallax and Triangles

The traditional method for calculating astronomical distances is the familiar one of calculating triangles from the knowledge of one side's length and two angles. If an observer sees the Moon overhead while another observer, standing on the first observer's horizon, measures his angle of sight to the

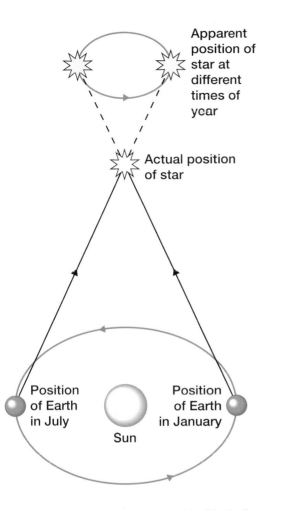

Apparent position of star at different times of year

Actual position of star

Position of Earth in July

Sun

Position of Earth in January

Moon, then the length of the other two sides—which will be the distance to the Moon—can be calculated. This technique works well for objects that are near Earth, but for distant stars a much longer baseline is needed. Astronomers learned how to use the orbit of the Earth around the Sun as that baseline: Taking observations of a star six months apart would give a baseline that was the diameter of the Earth's orbit, around 190,000 miles. The apparent shift in the star's position was called a parallax shift, exactly the same as the way your outstretched finger seems to jump as you close first one eye then the other. This technique reached its limit when the baseline became so thin in relation to the great distances of the stars that other techniques had to be found.

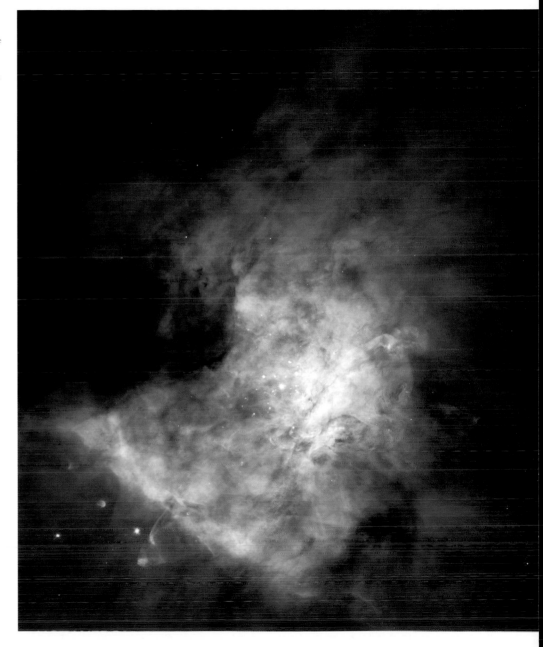

THE REDEFINITION OF SPACE AND TIME: EINSTEIN

In the new physics that took shape in the early twentieth century one scientist stands out because he revolutionized our understanding not of the microscopic nature of matter but of the macroscopic (visible) conditions under which matter forms the universe. In order to describe the mechanics of the universe, three basic factors are required: space, time, and mass. It was Albert Einstein (1879–1955) who showed that these three things are related to each other in ways that had never before been truly understood. Einstein was often pressed for a quick, nontechnical summary of this ideas, and on one occasion he replied by saying that before his theory of relativity, it had always been assumed that if all matter vanished from the universe, space and time would remain; according to relativity, he said, this was not true.

Analyzing Motion

Einstein's theories were published between 1905 and 1916. His center of interest was to analyze motion in different frames of reference. A ship on the sea, a passenger on a ship, the sea on the Earth's surface, the Earth itself, the solar system—all were in motion in different directions, and Einstein asked whether there was any absolute standard of time and space by which these movements could be related to each other. His answer was that there is not, because all events are defined by light, and light takes a finite time to travel, so that any event or any movement will appear to occur at a different point in time and space to observers in different frames of reference; there are in fact no simultaneous events.

Defining Light

Einstein's theories aroused worldwide interest, partly because they appeared so paradoxical, running counter to common sense; but it should be emphasized that the effects of relativity are not detectable in everyday experience. What Einstein did was to construct mathematical models involving scales of speed and distance far greater than those encountered in normal life and to calculate their effects. In this way he calculated that as speed increases, time moves more slowly, until finally, at the speed of light, time stops altogether. How can this be so? Because light defines all events: If something were moving at the speed of light, light could never catch up, and the event could never happen. This is one of the reasons why the speed of light is the top limiting velocity in the universe. The same effect is true at much lower speeds, those of a train, for example, but they are not detectable by normal means.

In the same way Einstein showed that physical mass increases with speed. In classical physics a body's mass is regarded as constant irrespective of its motion, but Einstein showed that this was untrue. If mass is defined as resistance to change of motion, then it is clear that a greater force is needed to move an object at a greater speed. At very high speed more and more force is required to increase the speed even further, until at theoretical speeds verging on that of light, infinite force would be required for any increase. Once more it was proved that nothing could exceed the speed of light. This aspect of relativity has been verified over and over again at the atomic level by studying the behavior of particles in accelerators. Infinitesimally small electrons at speeds approaching that of light gather enormous mass.

Energy and Mass

Following from this discovery, Einstein went on to make a deduction of huge importance both intellectually and practically. Since the mass of a moving body increases with motion, and since motion is a form of energy, the increased mass must come from the energy. Therefore, Einstein argued, energy has mass, or rather, energy and mass are interchangeable, and the distinction is one of temporary states.

By taking his calculations of the effects of increasing speed toward that of light, Einstein arrived at the celebrated formula $E = mc^2$, which showed the amount of energy concentrated into any physical mass: In any particle of matter the energy locked up is equal to the mass, in grams, multiplied by the square of the speed of light, in centimeters per second. This means that if one kilogram of coal were entirely transformed into energy, it would yield 25 billion kilowatt hours of electricity, enough to power a nation for weeks. This transformation could only occur by disintegrating the atomic nuclei and should not be confused with the chemical reaction that occurs in normal burning. But it must be remembered that when Einstein formulated this equation in 1905, neither he nor any one else had any clear idea of the structure of the atom or of the way in which its energy could be unlocked. Einstein was the father of nuclear power and nuclear weapons, but in an intellectual sense only.

Abiding Mystery

The convertibility of mass and energy provides part of the answer to some of the deepest puzzles in physics. It shows how radioactive substances such as uranium can send out charged particles for thousands of years. It explains

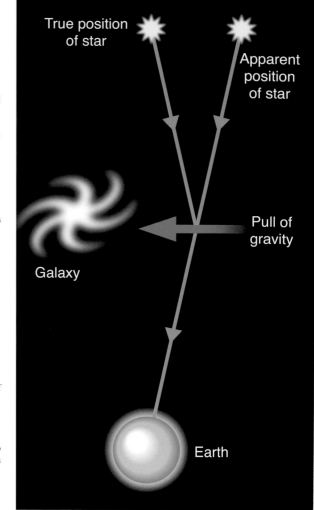

True position of star

Apparent position of star

Pull of gravity

Galaxy

Earth

how the Sun and other stars can go on radiating heat and light for billions of years. Perhaps above all, it illuminates the dual nature of matter, why sometimes it behaves like charges of radiation or electricity, but at other times it has measurable mass. If matter loses its mass and travels with the speed of light, it becomes radiation or energy. If that energy stabilizes into mass, it forms chemical elements, and we call it matter. Yet the mystery remains: What it "really is," where it originates, and what its ultimate end will be. The whole thrust of science in the nineteenth and twentieth centuries was toward the unification of nature's forms and forces: the conservation of matter, the conservation of energy, the laws of thermodynamics, and now Einstein's equivalence of matter and energy. All these fundamental laws seem to show that the universe is a single process: It is a succession of forms taken on by elements that are themselves eternal, moving in a vast, complex, and unending cycle.

Depths of the Cosmos

It is on the atomic and cosmological level that Einstein's ideas have been most profoundly influential. If space and time only exist relative to matter, then it no longer makes sense to separate them: Instead, they form a space-time continuum. That continuum is defined by light, and light, as mass-energy, should be subject to gravity—it should curve around matter. Gravity, in Einstein's view, was therefore the curving of space-time due to the presence of matter in the universe: If there were no mass-energy in the universe, there would be no space-time.

The bending of light in the gravity field of a massive star—the Sun—was verified in a famous experiment carried out in 1919 by the English astronomer Sir Arthur Eddington (1882–1994). This concrete proof of Einstein's bizarre and apparently abstract theories resulted in his being enveloped in a tidal wave of fame and hailed as the author of a new vision of the universe. The very strangeness of his ideas added to his celebrity, for example, when he announced that the universe must be considered finite but without limits. This concept was also proved true by the new cosmology that emerged between 1920 and 1950.

It was commonly said that Newton's physics had been overthrown, but Newton's physics still governs our terrestrial experience—for example, they are all that is needed for aerial navigation or for space travel within the solar system. Einstein's physics take us into the realms of the atom, the speed of light, and the depths of the cosmos.

The last 30 years of Einstein's life were spent in the search for a "unified field theory" that would embrace gravity and electromagnetism, the cosmos, and the atom. He was deeply distrustful of quantum physics and felt that it must be possible to build a model of physical reality as it really is and a mere probability system. His criticism of quantum theory—that "God does not play dice with the universe"—has become famous. Perhaps Niels Bohr's reply deserves equal weight, that "we cannot tell God how to organize the universe."

ABOVE RIGHT: Einstein with his daughter and son-in-law

RIGHT: At speeds approaching that of light, time slows and mass increases.

BELOW LEFT: Einstein's universe. The universe is composed of matter, space, and time, and Einstein explained how all three are inter-dependent. without matter there is no absolute space or time, in the same way a room is only defined by the four walls which enclose it. This means that space and time are fluid and have properties which can change under different conditions.

OPPOSITE PAGE:
RIGHT: Einstein predicted correctly that light would be bent by gravity, as part of his contention that mass and energy were interchangeable.

LEFT: Bohr and Einstein, the two most influential physicists of the twentieth century.

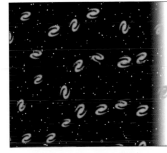

Nothingness
No matter = no time/space =
Nothingness

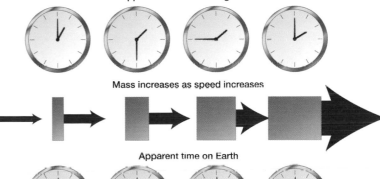

Apparent time on moving mass

Mass increases as speed increases

Apparent time on Earth

and he carefully calculated their periods and their luminosities, from which he estimated that they lay at least 100 million light-years away from Earth. This was far beyond Shapley's largest estimate for the dimensions of our galaxy and was in line with Curtis's ideas on the possible extent of the universe. Hubble's results were announced in December 1924, and it was generally recognized that a landmark in cosmology had been passed, and that a new scale and possibly a new structure for the universe had been revealed. The great question now was whether it was possible to determine how far the extragalactic universe extended and whether it could be studied.

The way forward clearly lay through a study of the galaxies themselves, their character and their distribution in space. How important were the different shapes that the galaxies displayed? Were they distributed at random through space, or was there some significant large-scale structure to the universe? Hubble played a leading role in this field, and over the following three decades he produced a classification system of galaxies showing the many variations on the basic elliptical and spiral forms, although he did not theorize about any possible process of evolution that might have led to these forms. His other fundamental task was to build up a scale of distances using natural markers such as the variable stars wherever possible.

By comparing types of stars in the distant galaxies, Hubble gradually built up an ever-growing picture of the cosmos, so that by 1930 his observations led him to believe that the visible objects that we see in the heavens extended over a distance of at least 250 million light-years. Hubble was aware that his estimates carried a degree of risk and uncertainty; but when the 200-inch Mount Palomar telescope was commissioned in 1949, his views were not only confirmed, but it appeared that he had underestimated his intergalactic distances by a factor of two.

Island Universes

The understanding of the galaxies as island universes permitted a new approach to the problem addressed by Hubble, that of mapping the universe in three dimensions. At first Hubble thought he had identified a large region in the sky where no galaxies appeared, but this was later explained as an optical aberration caused by dust in deep space. It then appeared that the distribution of galaxies was uniform throughout space when taken on a large scale. This observation was later extended by the English astronomer E.A. Milne into a "cosmological principle" that the universe is uniform in all directions. Later work on galactic mapping since Hubble's death has in fact revealed significant clusters or major groups of galaxies, which has raised a question about this principle of uniformity. Whether these clusters have formed as galaxies have drawn together is still undecided.

A Second Copernicus

Hubble's most historic discovery came in 1929, and it related to the spectra of the galaxies. Their light was found to be shifted to the red end of the

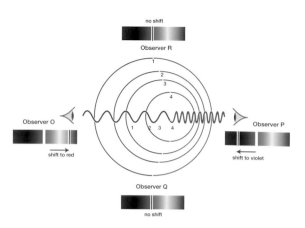

Unit 13 Red Shift

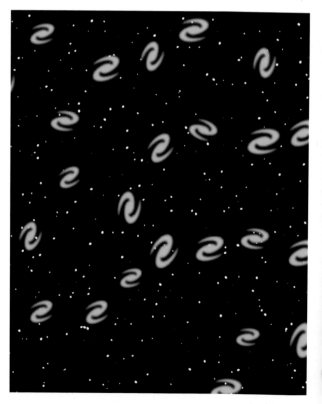

THE EXPANDING UNIVERSE: EDWIN HUBBLE

The impasse concerning the scale of the universe was resolved during the 1920s through the work of another astronomer who arrived at the Mount Wilson Observatory in 1919, Edwin Hubble (1889–1953). Using the new 100-inch reflecting telescope, Hubble looked for a way to settle the conflicting views of Shapley and Curtis, whether our galaxy was coterminous—having the same boundaries in time and space—with the universe, or whether many other similar star systems existed throughout an even wider space.

Landmark in Cosmology

The means to Hubble's decisive new discovery came through the same type of variable stars that Shapley and others had already studied. By 1923 Hubble had identified no fewer than 36 such stars in the Andromeda nebula,

FAR LEFT: Edwin Hubble in 1926.

LEFT: The Doppler effect: Light moving toward the viewer is squeezed into a shorter wavelength; light moving away is stretched into a longer wavelength. The red-shift is a crucial piece of evidence in our picture of the expanding universe.

BELOW LEFT: Island universes: Each galaxy is a self-contained system of stars, distanced by millions of light-years from the other systems; our own galaxy is but one island universe among countless others.

RIGHT: Portrait of American astronomer Dr Edwin P Hubble circa 1945 smoking a pipe.

BELOW: Hubble in 1948.

RIGHT: Hubble makes the cover of *Time* magazine February 9, 1948.

CENTER RIGHT: Hubble runs the 48-inch Schmidt Photographic Telescope through its final series of rehearsals for the National Geographic Society-Palomar Observatory sky survey. The project would provide the first ever definitive photo atlas of the heavens.

spectrum—the characteristic sign of a receding light source: Therefore these galaxies, which were already immensely distant, were actually moving away from Earth. Moreover, Hubble was able to show that there is a constant ratio between the distance and the velocity: The further away they were, the faster was their speed. Some of the galaxies observed by Hubble appeared to be moving at velocities of up to one-seventh of the speed of light. Nor was there any reason to suppose that this movement was only away from the Earth. The Earth could not possibly be imagined to be the center of the universe, and the cosmological principle suggested that the movement would be seen from any chosen point. The conclusion seemed inescapable that the galaxies were all rushing away from each other at enormous speed, and that the entire universe was therefore expanding.

This discovery was likened to a second Copernican revolution, for in place of an eternal, unchanging, motionless cosmos there now appeared a universe of intense, explosive movement. So strange was this discovery that even Hubble himself had some doubts about his findings. He wondered if some unexplained optical effects might be at work, so that the universe might be truly static. It is important to notice that everything in the universe is not expanding: The galaxies are not expanding internally, although they are in internal motion. The movement of the star Sirius that William Huggins found in 1868 was the motion of that star in relation to other stars and not part of the cosmic recession. The work of Hubble, building on the insights of Curtis and others, showed that it is the galaxies that are the large-scale units of the universe, and it is their distribution that is the key to the cosmic structure.

Big Bang: The Exploding Universe

Even before the extraordinary results of Hubble's observations became known, a number of physicists had been studying the effect of some of Einstein's ideas on cosmology. The Russian Aleksandr Friedmann (1888–1925) and the Belgian Georges Lemaître (1894–1966) argued that the universe must be nonstatic (in other words, constantly developing and changing) and that the curvature of space must be increasing with time. Hubble's work offered clear observational proof of this, and it raised two overwhelming questions: If the universe was really expanding, what had it expanded from, and what was it expanding into?

Expanding Sphere

To take the second question first, the relativity of space to matter and light could now be understood in concrete terms: As the galaxies moved into ever-more remote empty space, so their presence defined that space and with it the structure of the universe. There could be no absolute space divorced from the material bodies that mark it out. The paradox that the universe was at the same time finite but without boundaries now made sense, and so did the curvature of space, for the universe was now imagined to be an expanding sphere in which the distances between all points was simultaneously increasing.

TIME

THE WEEKLY NEWSMAGAZINE

ASTRONOMER HUBBLE
Will Palomar's 200-inch eye see an exploding universe?
(Science)

ABOVE: The "Big Bang" theory was developed in the 1940s and 1950s and dominates modern cosmological thought. The Big Bang, was the huge explosion that cosmologists believe created the universe. It occurred about 12–15 billion years ago, although the exact figure is uncertain. The theory is based on the fact that the visible universe is still expanding outward from a central point and on the discovery of background microwave radiation thought to be an afterglow of the explosion. This image from the HUDF-JD2 Hubble Ultra Deep Field shows one of the oldest galaxieswe have ever seen, one that grew within the first few hundred million years after the Big Bang—unlike our Milky Way that took billioins of years to reach its current size.

GEORGE GAMOW (1904–68

- Physicist.
- Born Odessa, Ukraine.
- Attended Leningrad University.
- Researched physics at various universities, including Göttingen, Copenhagen, Cambridge.
- 1931–34 Professor of physics at Leningrad University.
- Moved to the U.S. to become professor of physics at George Washington University (1934–55).
- 1948 With Ralph Alpher and Hans Bethe devised the "Big Bang" theory of the creation of the universe.
- Researched molecular biology, working on the order of nucleic acid bases in DNA chains. Made a major contribution to the understanding of DNA by discovering the "codes" within the proteins.
- Successfully wrote many books on popular science.
- 1956–68 Professor at Colorado University.

The surface of the Earth or any other sphere has the same property: It is finite but without boundaries, and the galaxies could be visualized as points on the surface of an expanding balloon. But if we take this expansion back in time, if we imagine the film being rewound, how had it begun? If matter is everywhere receding, it seemed logical to suppose that it was once much closer together, and Lemaître introduced the quantum concept into the discussion. If we go back in time, fewer and fewer quanta (units) of energy will be found, until the whole universe is packed into a single quantum, an atom of unimaginable density. In this single point, Lemaître suggested, some instability had arisen causing an immense explosion that was the starting point of the expanding universe. This was the first appearance of the Big Bang theory of creation, although Lemaître was not able to give it any precise mathematical form or to suggest a cosmic time scale.

Modern Myth?

A much fuller and more detailed theory of the mechanics of the Big Bang came from the Russian-born George Gamow (1904-1968) in 1948, in a paper that has become famous in scientific history as the "Alpha-Beta-

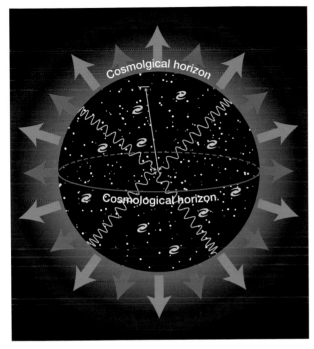

61

Gamma" paper because its coauthors were Ralph Alpher and Hans Bethe. Gamow himself later wrote a fuller account in his book *The Creation of the Universe* (1952). He suggested that all the protons, neutrons, and electrons in the universe had come into being at the moment of creation at unimaginably high temperatures, and that the structurally simple elements hydrogen and helium had been formed at a very early stage by nuclear aggregation. The heavier elements had, he believed, been formed later in the interior of stars, a view that later came to be universally accepted. Gamow tried to place the expansion of the universe within a time frame, calculating that some 17 billion years would be needed for the universe to reach its present size and shape. He made a prediction of the greatest importance, namely, that the radiation from the initial explosion should still be present permeating the universe even after such a vast passage of time.

The Big Bang theory dominates modern cosmological thought; indeed, it can be seen as the modern equivalent of old creation myths. Like the ancient myths, it is an explanation only in a limited sense: It offers a narrative of certain events, events that show the operation of some transcendent power. It satisfies the need for some explanation of the overwhelming mystery of how the universe has come to exist; but the ultimate how and why of these events are still beyond the reach of understanding. In the ancient myths it was a

god who shaped Earth and stars out of chaos; but where did the god and the chaos come from? In modern thought it is the laws of physics that have shaped the universe of matter, but who formulated those laws? The only rational, nonreligious answer is that they are inherent in matter itself. If matter is to exist, it must exist in an ordered state. But why should matter exist at all? There is always another question that can be asked that neither science nor myth can answer. It is interesting to see that many astronomers and physicists have once again started to speak of God, although using the word in a quite impersonal way. They speak of God as the unfathomable power that arranged the universe from its birth, and that has determined what the laws of physics are. This impersonal God is proving useful as a

OPPOSITE, LEFT: Hubble reading a journal at Mt. Wilson Observatory.

OPPOSITE, RIGHT: November 1937: Hubble looking though the eyepiece of the 100-inch telescope at the Mt. Wilson Observatory.

ABOVE: Philip Staples (left), President of the Franklin Institute, presenting the Franklin medal to Edwin Hubble, of The Mount Wilson Observatory, Carnegie Institute of Washington, Pasadena, California, in recognition of his extensive study of the Nebulae. The presentation was made during the Franklin Institute medal day ceremonies, May 17, 1939.

ABOVE LEFT: Light coming from the horizon of the universe (red arrows) is red-shifted to infinity because those galaxies are receding; as the universe expands, the horizon expands even faster (blue arrows), bringing new sights into view.

summary of all that cannot be explained or rationalized any further, but just is.

The Elements of the Universe: B2FH

A further factor entered into astronomers' attempts to map the universe with the discovery of the interstellar medium—that the space between the stars and the galaxies is not empty but is filled with matter in the form of gases and very diffuse particles. In the late 1920s Robert Trumpler (1886–1956), a Swiss-born American astronomer, became puzzled by discrepancies in the size, brightness, and therefore the apparent distances of star clusters that he was studying. He conceived the idea that their light was being absorbed by an invisible medium, and that it was distorting our perception of them. In one particular area of the night sky, a strip about 20 degrees wide centered on the Milky Way, no nebulas at all were visible, so that it was at first believed by Hubble and others that no external galaxies existed in this "zone of avoidance." Trumpler understood correctly that the effects of this medium increased with distance, so that farther objects would appear fainter, leading to an overestimate of their distance. Those astronomers such as Harlow Shapley who had made estimates of the size of our galaxy accepted the validity of Trumpler's discovery and revised their figures downward by a large amount. Although the interstellar medium was very diffuse, it was clear that it must contain a huge volume of matter.

Space Dust

The dust in space was far from being a negative discovery, for it provided an answer to the problem of the birth of the stars. Hertzsprung and Russell had concluded that stars evolve over time. Next, Eddington and later Hans Bethe showed how this might happen on the atomic level. But where did the matter come from from which the stars evolved? Some stars were very old, and some were much younger, suggesting that there was a "life cycle" and a source of their material. Some years before the work of Hubble, before it had been decided that nebulas were star systems, the strange forms of the spiral-armed galaxies had attracted much attention: How could they possibly have arisen?

In 1919 the Cambridge physicist James Jeans (1877–1946) explained in theoretical terms that a spherical mass of gas would contract under the force of gravity, then start to spin until it became unstable. It would then throw out filaments of matter from its edges, which could then condense into spirals, exactly as seen in these nebulas. When huge volumes of gas were found to exist in interstellar space, added force was given to this theory, which in general terms is still believed to be correct. It was then discovered that where the interstellar medium was densest, it could be subject to spectral analysis, showing that it contained light gases, mainly hydrogen, but also nitrogen, oxygen, and dust particles: The interstellar space was found to be full of huge quantities of basic chemical elements. The gases were not evenly distributed, but were to be found in patches, some relatively dense. On the principle outlined by Jeans,

it seemed that the birth of stars would occur when these gases condensed under the force of gravity and became sufficiently hot to start nuclear reactions.

Evolutionary Theories

From the 1930s onward theories concerning the evolution of stars were guided by a deepening understanding of atomic physics as observed here on Earth. That elements could transform themselves one into another by losing or capturing protons and neutrons was well known, but it applied to the lighter elements only. The process of star formation could not apparently account for the existence of all the heavier elements that are found on Earth, and that had been found in Sun's spectrum as long ago as the 1860s by Kirchhoff and Bunsen.

This problem was studied by many leading astrophysicists, and in 1957 there appeared a paper entitled *"The Synthesis of Elements in the Stars"* written jointly by four scientists, Margaret Burbidge, Geoffrey Burbidge, William Fowler, and Fred Hoyle. This historic paper has therefore become known in the scientific world as B2FH, and it set out the physical rules for the creation of the elements past hydrogen and helium, which, its authors argued, was taking place continuously inside stars. The nuclear reactions inside the stars

ABOVE: August 20, 1941, Los Angeles. Hubble is at far left.

LEFT: January 16, 1931, in Pasedena California, scientists meeting Prof Albert Einstein. L–R: Dr M. L. Humason; Edwin Hubble; Dr Charles Saint John, Albert A. Michelson; Prof Albert Einstein; Dr W. W. Campbell; W.S Adams.

B2FH (GEOFFREY AND MARGARET BURBIDGE, WILLIAM FOWLER, FRED HOYLE)

Geoffrey Burbidge (1925–)
- Astrophysicist.
- Born Chipping Norton, England.
- Attended Bristol University, then University College, London.
- 1950s Collaborated with Fred Hoyle on the astrophysical consequences of antimatter.
- 1951 Moved to the U.S., first to Harvard, then Chicago, Mount Wilson, Palomar, and California.
- 1957 In collaboration with his wife Margaret, Hoyle, and William Fowler published a paper of theoretical research on nucleosynthesis—nuclear physics as applied to astrophysical circumstances.
- 1967 With Margaret published an important work on quasars.
- 1970 Highlighted the "missing matter" problem of galaxies.

continued on page 64

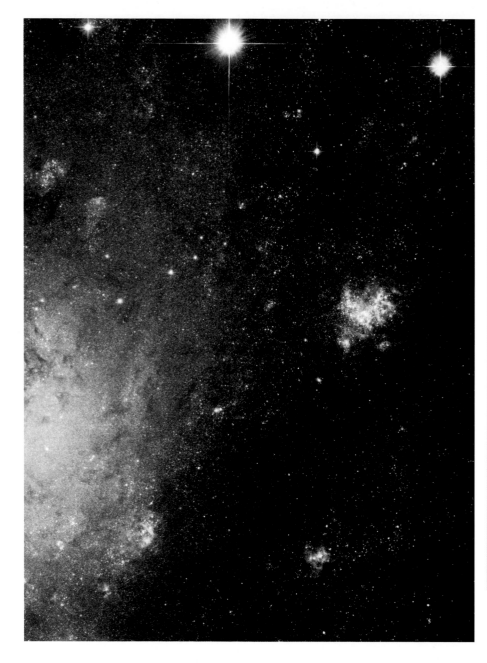

Margaret Burbidge, née Peachey (1923–)

- Astronomer specializing in galaxies and quasars.
- Born Davenport, England.
- Attended University College, London, and became interested in astronomical spectroscopy.
- 1948–51 Assistant director of London University observatory.
- 1951 Moved to the U.S. to Yerkes Observatory, then Caltech, then to the University of California.
- 1964–90 Professor of astronomy, University of California.
- 1972 Worked as director of the Royal Greenwich Observatory for a year
- 1979–88 Director of the Center for Astrophysics and Space Science at San Diego.

William Fowler (1911–95)

- Physicist and founder of the theory of nucleosynthesis.
- Born Pittsburgh, Pennsylvania.
- Studied at Ohio State University and California Institute of Technology.
- 1936 PhD from Caltech for work on radioactive nuclides.
- 1946 Appointed professor at Caltech
- Work concerned measuring nuclear reactions at low energies. Established the existence of the postulated excited helium state.
- After 1957 B2FH continued working on stellar nucleosynthesis and solar neutrino flux calculations.
- 1983 Shared the Nobel Prize for physics with Subrahmanyan Chandrasekhar.

Fred Hoyle (1915–)

- Born Bingley, England.
- Schooled at Bingley Grammar School and Emmanuel College, Cambridge.
- 1945–58 Taught mathematics at Cambridge.
- 1948 He and two colleagues advocated the "steady state" of the universe, that the universe is uniform in space and unchanging in time—a now discredited theory.
- Worked on investigating supernovas and the re-cycling of second generation stars from the exploded matter of earlier stars.
- Plumian Professor of Astronomy and Experimental Philosophy at Cambridge between 1958 and 1972.
- A prolific author of fiction and scientific works.
- 1972–78 Professor-at-large for Cornell University.

were constantly releasing free neutrons, which were built into ever heavier elements, a process that they plotted in precise mathematical order. The process continues as while the star is active; but toward the end of its life it will collapse when its material is exhausted, and it will explode in one of several ways.

The most spectacular and visible way is for it to become a nova or a supernova, a bright, violently exploding star, such as the one that Tycho watched in 1572. What is happening in a nova is that the material of a star, the chemical elements, is being ejected in atomic form back into the diffuse interstellar medium, where the cycle of star creation can begin once again. The B2FH paper was a landmark not only in astrophysics, but in our whole understanding of the universe, for it showed that a vast physical cycle is at work in the stars that has produced all the chemical elements that constitute our Earth and everything on it, including ourselves.

Problems and Conflicts in Cosmology

The idea that the stars are engaged in a vast cosmic process, recycling the chemical elements in a life cycle extending over billions of years, inevitably appears to raise questions about the Big Bang theory of the universe. The authors of "B2FH," and in particular Fred Hoyle, were never reconciled to the Big Bang model; indeed, it was Hoyle himself who coined the phrase Big

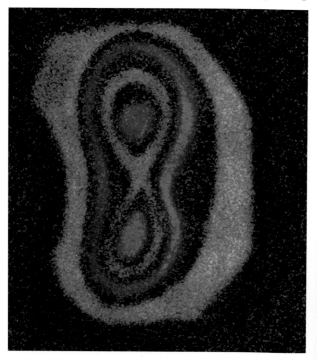

Bang as a joke, although it was subsequently accepted by the whole scientific community. Instead, Hoyle, working with two Austrian physicists, Thomas Gold (1920–2004) and Hermann Bondi (1919–) , devised the "steady state" concept of the universe—that matter was being continually created in interstellar space. They were not able to explain precisely how this happened, but as they pointed out, neither could the Big Bang theory; and it was surely no more extravagant to claim that matter was being continuously created than that it had all come into being at one instant.

Steady State

According to the steady state theory, the universe has no beginning and may have no end; it therefore removed some of the philosophical difficulties of the Big Bang, above all the problems of imagining what had existed before the moment of creation, and what had caused it. They argued that the Big Bang envisaged a time when the laws of physics did not apply, and then an arbitrary moment when they came, inexplicably, into existence, which they found absurd. There were several grave difficulties with the steady state model, however: If matter was being created, then the birth of galaxies, the building blocks of the universe, should be observable somewhere; but it was not. Nor could the dramatic recession movement of the galaxies be

explained. But the event that undermined the steady state theory came from the new technique of radio astronomy.

Radio Waves

Radio astronomy had emerged by accident in the 1930s, when Karl Jansky (1905–50), an engineer for Bell Laboratories, was asked to investigate interference that was hampering transatlantic radio signals. Jansky built a large antenna and, turning it to the sky, found constant interference with a recognizable pattern over a period of 24 hours. Jansky realized that it was the Earth's rotation that was imposing this pattern, and that the radio waves must come from outside the Earth: It was the stars that were constantly bombarding the Earth with invisible radio signals, just as they emitted visible light.

After World War II this technique was developed to analyze stellar radiation and build up a map of celestial objects quite independent of optical telescopes. Radio telescopes were the means by which cosmic background radiation was discovered. George Gamow had predicted that a radiation echo from the Big Bang should still be found permeating the universe, and in 1965 Arno Penzias (1933–) and Robert Wilson (1936–), two radio astronomers, found that every part of the sky, even a seemingly empty region, is giving off short-wave radiation at a constant low temperature of

slightly under 3° Kelvin. This radiation was entirely even and uniform, not associated with individual objects, but permeating the whole sky. It was accepted as being the echo of the Big Bang, which Gamow had predicted, and the steady state theory had few supporters after this discovery.

Strange Characteristics

Radio telescopes have been responsible for broadening our knowledge of the universe far beyond the visible stars. In 1967 the first pulsar was identified by Jocelyn Bell (1943–) at Cambridge in England. It was detected as a fast-flashing radio beacon of a kind so novel that at first it was thought to be a communication from beings in deep space. Soon others were found, however, and it was concluded that pulsars were the remains of collapsed stars only a few miles in diameter, yet more massive than the Sun and spinning wildly, sometimes several times per second. Pulsars can be seen with optical telescopes, but their strange characteristics could never have been identified except through their radio signals.

Even stranger are the quasars, first detected through radio emissions, but later located optically. Quasars are bodies that are no more than a light-year in size, but have a tremendous luminosity equal to that of a whole galaxy with a diameter of 100,000 light-years. This permits them to be seen at

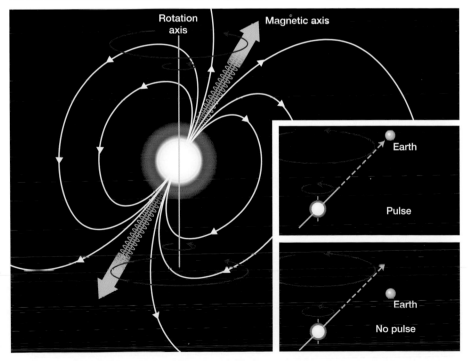

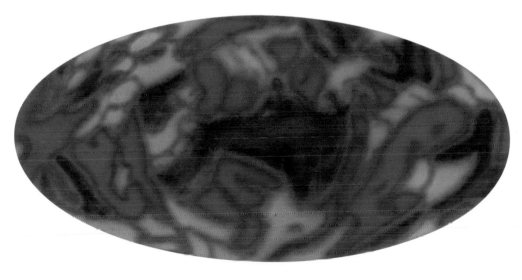

-0.27   +0.27

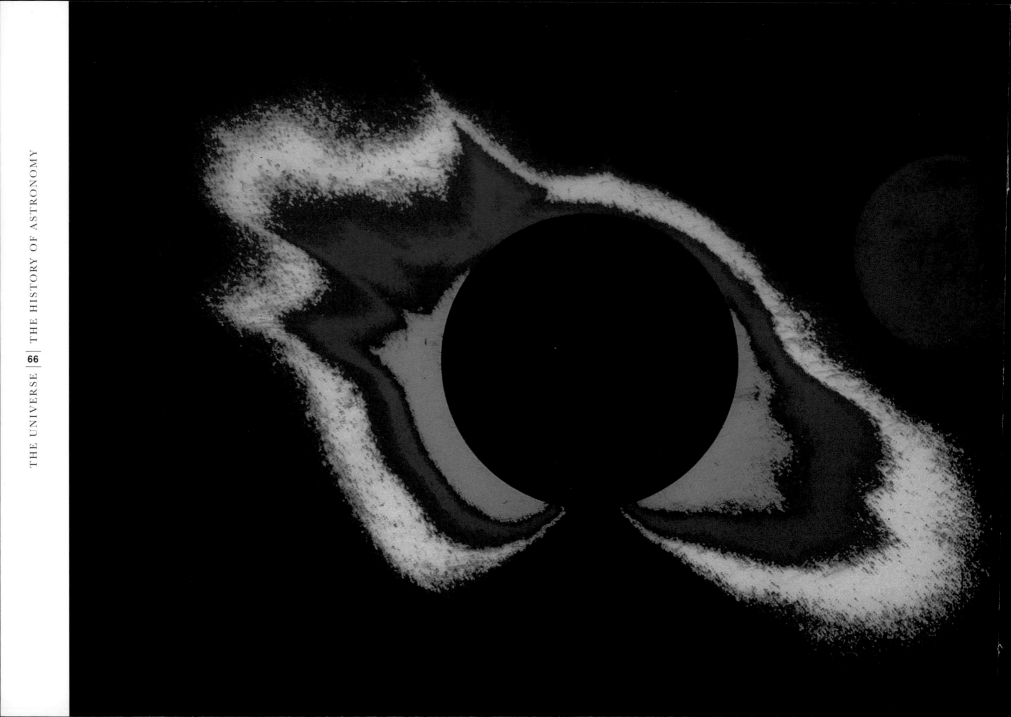

cosmic distances of more than one billion light-years; in fact, the most striking thing about quasars is that they do not seem to exist at less than about three billion light-years distance, and most are much further away.

At the cosmic scale the viewing of distant objects is also a measure of time: When we see a quasar that is ten billion light-years away, we are seeing an object as it existed ten billion years ago. The clear implication is that quasars may not exist any longer, that they were an intermediate stage in cosmic evolution, and many astronomers believe that they are embryo galaxies.

In present-day astronomy the hunt is on for perhaps the strangest object known to astrophysics, the black hole. It has been suggested that in certain circumstances a star may collapse to such a dense state that no radiation, not even light, can escape from its surface. Matter approaching this dark region will be sucked into it by gravity and become invisible. For obvious reasons no black hole has ever been positively seen, but certain points of high radiation turbulence may be the edges of black holes. They are of great interest to physicists because they may represent areas of the universe where the laws of physics do not apply and may therefore hold the key to the nature of the universe before the Big Bang.

Big Squeeze?

The age of the universe and the size of the universe are two interrelated questions. In the mid 1930s, after the meaning of Hubble's discovery had become clear, a size of two billion light-years for the universe was suggested by Englishman Arthur Eddington (1882–1944). But work on the evolution of stars and galaxies implied that their individual ages were greater than that. The advent of radio telescopes and the opening of the giant 200-inch optical telescope at Mount Palomar in 1949 both revealed galaxies at greater distances than two billion light-years, so once again the scale was extended. The current estimate is around 15 billion light-years in extent and 15 billion years for the age.

The two are inseparable, because when we see across 15 billion light-years of space, what we see is light or radiation that left its source 15 billion years ago. But there is no real reason to believe that this is a final figure, since cosmological thought is still very fluid. Astronomers cannot agree on the central question—whether the universe will continue to expand forever, or whether it will one day slow, halt, and go into reverse-collapse—the so-called big crunch or big squeeze. There is a tendency among today's astrophysicists to speak as though many of these fundamental problems were on the verge of being answered or had even been answered already. It should be remembered that ideas such as the Big Bang are still only theories, however widely they may be believed at present. It may be that further Copernican revolutions in our understanding of the universe are waiting to happen in the future.

The Birth of the Solar System

In 1906 an American astronomer and a geologist, F.R. Moulton (1872–1952) and Thomas Chamberlin (1843–1928), announced a startling new theory of the origin of the solar system. They suggested that at some time in the remote past, a second star had wandered close to the Sun, and that its gravitational pull had dragged a stream of matter out of the Sun, and the stream had then cooled to form the planets. This was a radically new idea, for until that time the nebular hypothesis of Laplace had dominated scientific thought. The Moulton-Chamberlin theory was widely believed, and one of its implications was that the solar system, having arisen from an extraordinary and perhaps unique event, might itself be unique.

In the 1920s and 1930s eminent scientists such as James Jeans and Arthur Eddington considered that this might indeed be true, for no other planetary systems had been observed surrounding other stars. This is still true, and the history of the solar system is still a matter of theory only: Strangely, modern astronomers seem more certain about the origin of the universe as a whole than of our own planet and its immediate neighbors. In the nineteenth century geologists and naturalists such as Lyell and Darwin had argued that the age of the Earth must be measured in hundreds of millions of years. This conflicted with the best estimates of physicists like Lord Kelvin that the Sun itself could be no more than a few million years old, otherwise it would burn itself out.

Meteorite Samples

The discovery of nuclear energy as the source of the Sun's power solved this problem, while techniques of dating through measuring radioactive decay enabled Earth scientists in the 1960s to estimate the age of the Earth as around 4.5 billion years. This figure was confirmed when geologists were able to analyze meteorite samples and then lunar samples. Astrophysicists also agreed that the Sun is approximately halfway through its life cycle of 10 billion years: It seemed therefore that the entire solar system had originated at the same time. The Moulton-Chamberlin theory was generally rejected, and scientists turned back to the earlier nebular model. Since we know from spectroscopy that all the heavier elements that make up the Earth—iron, magnesium, carbon, and so on—are also present in the Sun, it seemed that Laplace had been right after all: Gases had condensed to form the Sun and the planets at the same time.

Solid Particles

The nebular hypothesis was revived by the German Carl von Weizsäcker (1912–) in 1945. Beginning with the protosun surrounded by a disk of rotating gas, Weizsäcker argued this mass would break up into smaller vortices or eddies, which in turn condensed to form the planets. The differing composition of the planets, from the small, solid, rocky, iron-cored inner planets, to the massive but very cold gaseous outer planets, was

PREVIOUS PAGES:
PAGE 64
LEFT: A supernova (the bright star at top right of image) an exploding star that returns its chemical elements to interstellar space, to be recreated into new stars.

RIGHT: The quasar appears to distort the light from a nearby galaxy and its red shift is identical to that of the galaxy, showing that it lies at a cosmological distance.

PAGE 65
LEFT: A pulsar's magnetic field spins in space around its axis of rotation. Radiation is beamed along the magnetic axis and will strike the Earth if suitably positioned.

RIGHT: Cosmic background radiation, a crucial piece of evidence for the Big Bang theory; the red areas are hotter, the blue cooler, but the distribution is uniform.

LEFT: False color image of the Sun showing the corona.

determined by temperature differences as the vortices were spread further from the Sun.

Subsequent theories gave a greater role to the solid particles that are found in the interstellar medium, ice-covered particles of silicon, carbon, and other elements. It is now believed that they came together under the force of the protosun's gravity, forming "planetesimals," which grew in size until they formed the planets. The asteroids and other bodies in the solar system—including some of the satellites of Jupiter and perhaps even the planet Pluto—may be large surviving planetesimals. The dynamics and the chemistry of this process are still the subject of theoretical investigation.

From time to time astronomers have suggested new catastrophic theories as playing a part in the history of the solar system, for example, that the Moon was split from the Earth by a collision with another large body. Unlikely as this sounds, it would explain some geological similarities between the two bodies. In complete contrast, it has also been argued that the Moon was originally a body wandering from a distant part of the solar system, which was "captured" by the Earth's gravity.

The origin of the moons of Jupiter, Saturn, and Uranus are even more mysterious, some rocky, some icy, some volcanic, and many of them so different from their parent planets. If the origin of the solar system were more certain, we would be better able to say how likely it is that other such systems exist. It is a paradox of modern astronomy that despite penetrating the depths of space, our knowledge of the birth of our own solar system, and its possible uniqueness, is still so uncertain.

The Earth's Magnetism: James van Allen

The existence of the magnetic force is one of the oldest scientific facts known to humans, and it has been one of the most practical, for the invention of the magnetic compass in the late Middle Ages revolutionized the art of navigation. But what was the source of this magnetic effect? One of the earliest theories was that the compass was pointing toward the pole of the sky, until the Elizabethan scientist William Gilbert announced his discovery that the Earth itself was a vast magnet with twin poles. This idea was universally accepted, but for hundreds of years no scientific explanation of the Earth's magnetism was forthcoming until more precise methods of geophysical measurement were developed in the early twentieth century.

Shock Waves

Between 1895 and 1910 the English seismologist Richard Oldham (1858–1936) made careful studies of earthquakes in India, where he was the head of the Geological Survey. By comparing data from different monitoring stations, Oldham found that the seismic waves appeared to travel at varying speeds to places opposite the focal point of the quake. From this he deduced that the Earth's interior was not uniform, but that it contained a central core that was denser and more elastic, and through which the waves therefore

1. As the star forms, matter settles into a disk around it

2. Matter in the disk begins to clump together

3. Rocks form around the star

4. Planets like the Earth coalesce out of the rocky material

traveled faster. In later years the study of the speeds and paths of these shock waves through the Earth has made it possible to map the Earth's interior, showing its various solid and liquid layers. The existence of these layers has also provided the explanation for the Earth's magnetism, namely, that the Earth acts as a geomagnetic dynamo. Fluid motions in the liquid outer core, which consists largely of molten iron at high pressure and high temperature, set up an electric current, with its associated magnetic field; these motions are continually sustained by the Earth's rotation.

Van Allen Belts

The Earth's rotation and therefore its magnetic field have far-reaching effects on climate and therefore on life; indeed, in the region around this planet magnetism creates a zone known as the magnetosphere in which the magnetic field extends some 40,000 miles into space. In this zone the Earth's

RIGHT: False color image of Callisto, one of the 16 satellites of Jupiter and one of the four so-called "Galileans"—the others are Io, Europa, and Gannymede. They are called this because they were discovered and observed by Galileo in January 1610. He called them the "Medicean planets" in honor of the great Italian Medici family.

FAR RIGHT: The origin of the Moon is still unknown: Was it formed at the same time as the Earth, or was it torn from the Earth by a cosmic collision?

ABOVE: The modern theory of planetary origins: Interstellar matter coalesces (bonds) under its own gravity, and the cooler, outer regions form planets.

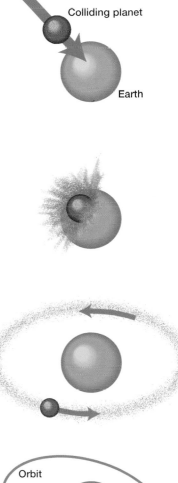

Colliding planet

Earth

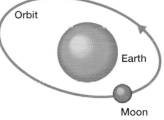

Orbit

Earth

Moon

magnetic force deflects most of the charged particles of radiation that come from the Sun. In the direction of the Earth's travel the magnetosphere resembles the bow wave of a ship moving through water, while behind it the tail stretches out much further beyond the Moon's orbit. The most significant regions within the magnetosphere are the two van Allen belts, identified by the American physicist James van Allen in 1958.

Van Allen (born 1914) was a naval scientist who transferred to rocket research after the end of World War II. A series of experiments carried out from the earliest orbiting satellites showed that a considerable amount of solar radiation leaked into the magnetosphere and became concentrated in two large, paired, crescent-shaped zones, one inner and one outer, that curved down toward the Earth's magnetic poles. Their function was confirmed when several small nuclear explosions were set off 300 miles above the Earth in order to release very energetic particles into the magnetosphere. Monitoring of this radiation showed that it had indeed been captured in the van Allen belts.

The Earth's magnetic fields, working through the van Allen belts, clearly protects the Earth from harmful radiation from space. We cannot say that this is its purpose, but it provides a striking example of balance and symmetry in nature: If this one aspect of the physical world were slightly different, life on Earth would be impossible.

WHAT REMAINS TO BE DISCOVERED?

The uncertainty about the origin of the solar system is a sobering reminder of the puzzles and mysteries that still await answers in the realm of astronomy. There has been a century of extraordinary discovery about the stars and the cosmos, but every discovery seems to lead to the unexpected or even to the inexplicable. Claims that we are on the verge of final answers about the origin and workings of the universe need to be treated with a great deal of skepticism: If astronomers cannot explain where our own planet and its immediate neighbors came from, can we really believe their accounts of the birth of the entire universe? There is a cluster of fundamental questions in cosmology that are currently impossible to answer, and that prevent us even now from forming a reliable picture of our universe.

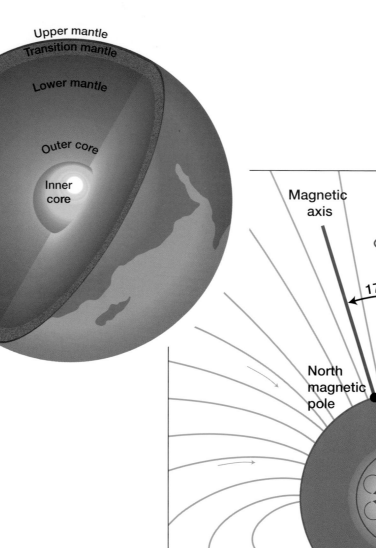

Upper mantle
Transition mantle
Lower mantle
Outer core
Inner core

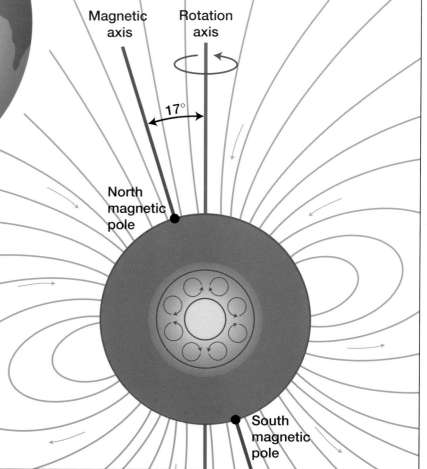

Magnetic axis   Rotation axis

17°

North magnetic pole

South magnetic pole

Older Universe?

First, there is Hubble's constant, the relation between the velocities of the remote galaxies and their distance from us. Hubble's constant should express the rate at which the universe is expanding; but what is Hubble's constant? Hubble himself estimated that it was 93 miles per second per million light-years, from which Hubble deduced that the universe was some 2 billion years old. However, modern cosmologist have revised this drastically downward, to between 9 and 18 miles per second per million light-years. This slower rate means that the universe must be much older, possibly 10 billion to 20 billion years; but this is a huge range, and current observations cannot make it any more precise.

Origin of Galaxies

Then there is the question of the galaxies, the basic building blocks of the universe: Do stars come together into galaxies under the influence of their gravity, or are galaxies formed as communities of stars out of vast clouds of interstellar dust? The former seems unlikely, for no stars have ever been observed that are not part of a galaxy; yet the process of the origin of galaxies is not observable either. Meanwhile, the distribution of the galaxies is still unmapped. Within a certain range of vision a number of major clusters have emerged, but no symmetrical pattern: Does the universe have a large-scale structure, or are the galaxies scattered randomly through space?

Expanding Universe

The most intriguing problem is whether the universe will continue to expand or not. Will the explosive force of the Big Bang one day exhaust itself and be defeated by the gravity between the galaxies, so that its expansion will slow and then halt, and all the matter in the universe will collapse back into itself?

Cosmologists claim that, in theory, they should be able to answer that question by comparing the mass of the universe with the velocity at which galaxies are moving apart. This calculation has proved impossible, however, mainly because of the "dark matter" that is now believed to fill space.

As long ago as the 1930s the Swiss astronomer Fritz Zwicky (1898–1974) tried to calculate the mass of a galaxy from its luminosity, using the Sun as a

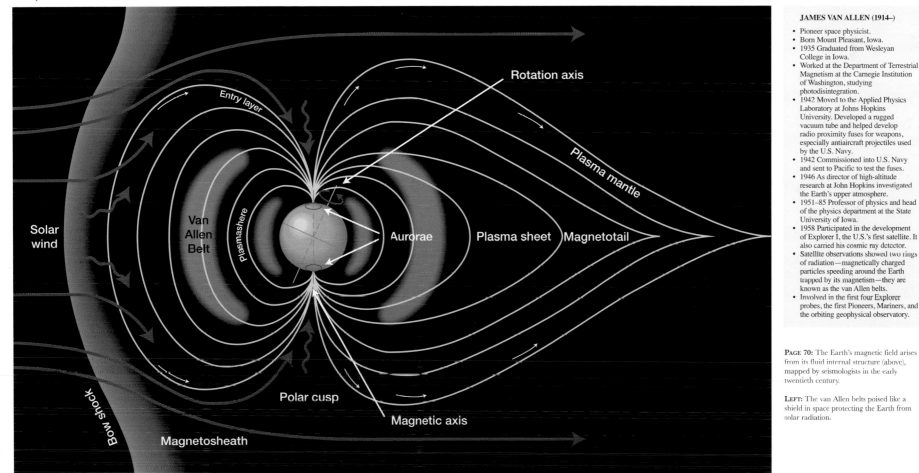

JAMES VAN ALLEN (1914–)

- Pioneer space physicist.
- Born Mount Pleasant, Iowa.
- 1935 Graduated from Wesleyan College in Iowa.
- Worked at the Department of Terrestrial Magnetism at the Carnegie Institution of Washington, studying photodisintegration.
- 1942 Moved to the Applied Physics Laboratory at Johns Hopkins University. Developed a rugged vacuum tube and helped develop radio proximity fuses for weapons, especially antiaircraft projectiles used by the U.S. Navy.
- 1942 Commissioned into U.S. Navy and sent to Pacific to test the fuses.
- 1946 As director of high-altitude research at John Hopkins investigated the Earth's upper atmosphere.
- 1951–85 Professor of physics and head of the physics department at the State University of Iowa.
- 1958 Participated in the development of Explorer I, the U.S.'s first satellite. It also carried his cosmic ray detector.
- Satellite observations showed two rings of radiation—magnetically charged particles speeding around the Earth trapped by its magnetism—they are known as the van Allen belts.
- Involved in the first four Explorer probes, the first Pioneers, Mariners, and the orbiting geophysical observatory.

PAGE 70: The Earth's magnetic field arises from its fluid internal structure (above), mapped by seismologists in the early twentieth century.

LEFT: The van Allen belts poised like a shield in space protecting the Earth from solar radiation.

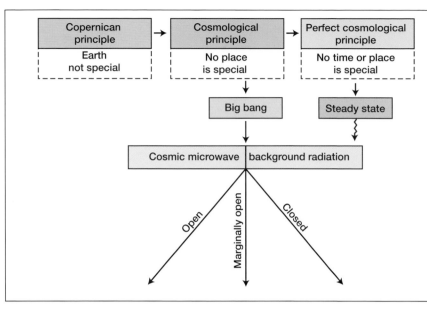

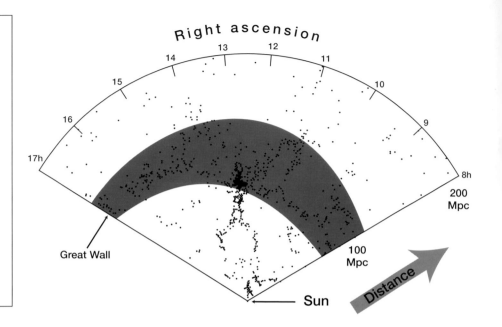

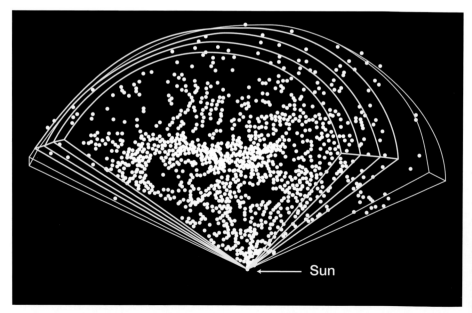

standard measurement. Zwicky was astonished to find that galaxies appeared to possess 50 times more mass than their luminosity would require, and this result was confirmed by other observers too. The only conclusion to be drawn was that galaxies were filled with matter so diffuse as to be invisible. This major unknown makes it impossible to say if the universe is "open" or "closed"—whether it will go on expanding forever or will one day contract into the dense state that existed before the Big Bang. This may seem a totally academic question, since in either case the end of the universe lies billions of years in the future, yet it is a burning issue in science because of humankind's consuming need to map the universe it lives in. But it is clear that the unknown in the universe still enormously outweighs the known. It will be time to put forward grand "theories of everything" when we know what everything is.

ABOVE: People's understanding of the universe has been revolutionized several times; the Big Bang theory may not be the last word, and its long-term fate is totally unknown.

RIGHT AND ABOVE RIGHT: Is there a large-scale structure to the universe? Plots of galaxy positions reveal strange concentrations, but no clear pattern.

FAR RIGHT: Mars taken through the Hubble space telescope in 1997.

The Hubble Space Telescope

The launch of the Hubble Space Telescope (HST) on April 24, 1990, aboard the space shuttle *Discovery* (STS-31) opened a new era in optical astronomy. Observers were now able to view objects in deep space unaffected by the distortions caused by the Earth's atmosphere. The telescope was named after the famous American astronomer Edwin P. Hubble who died at the age of 64 in 1953. His painstaking observations led to the confirmation of the concept of a constantly expanding universe which in turn led to the widely accepted Big Bang theory of its formation.

The HST lies in a low Earth orbit at an approximate altitude of 380 miles and completes one orbit every 97 minutes. Its complex mechanisms are powered by six nickel-hydrogen batteries which are constantly recharged by two 25ft solar panels. Almost immediately after it started returning images to Earth, it was realized that all was not quite right and an unacceptable fuzziness was apparent in some circumstances. This caused some uninformed critics to write the whole project off as an expensive failure but in fact the problem was caused by a very slight flaw in the telescope's primary mirror. In normal circumstances this would not have mattered but it is an indication of the overall accuracy of the telescope that an aberration equal to less than one-fiftieth the thickness of a human hair was noticeable. Although many of the images produced were still better that those produced by ground based telescopes, and an imaginative repair scheme was put in place. When the telescopes first planned servicing mission was launched in December 1993, it included device which placed pairs of corrective mirrors in front of the various cameras and other instruments. The mission was a complete success

and since then the HST has more than fulfilled the hopes of its builders and designers.

Other servicing missions have been flown, one in 1997 and another two-part mission flown in December 1999 and February 2002, in which various components such as batteries and gyros were replaced while existing equipment was upgraded and improved. Following the *Columbia* shuttle disaster in February 2003, another servicing mission was cancelled and consideration was given to launching a robotic module to effect essential repairs and replacements. However, such is the importance and standing of the Hubble Telescope that it has been accorded a high priority aboard future shuttle missions following recommencement in July 2005 .

There is no doubt that some of the images produced by the HST are absolutely spectacular and have awakened a new interest in the science of astronomy. In many of these, galaxies and nebulas appear as colorful masses making imaginary shapes such as the Horse's Head Nebula. In fact, viewed by the human eye through a conventional telescope, the vivid colors would not be apparent and, indeed, many of the features on the Hubble images would not be visible at all. The reason for this is that the HST observes and records not only the visible spectrum but also the ultraviolet and infrared spectrums. All of these are basically recorded in a monochrome grey scale but digital enhancement and the use of various filters can manipulate the image. The resultant colors enable new details to be detected and various measurements to be taken. The staggering artistic effects are something of a bonus, but none the less impressive for all that.

RIGHT: A view of the Hubble telescope as it drifts in orbit over 300 miles above the Earth. This view was taken during the last routine servicing mission in February 2002. The twin solar panels are clearly visible.

LEFT: *Discovery* blasts off from the Cape Kennedy Space Center on April 24, 1990, carrying the Hubble telescope into orbit. This represented the fruition of project which began as far back as 1969 when the idea of a large space telescope was formally accepted and feasibility studies began.

FAR LEFT: The Hubble Space Telescope rests in the cargo bay of the Space Shuttle *Discovery* during the December 1999 repair mission. It is attached to the shuttle during the maintenance period to enable astronauts to carry out necessary work Coincidentally, *Discovery* was the shuttle which originally placed the Hubble telescope in orbit in 1990.

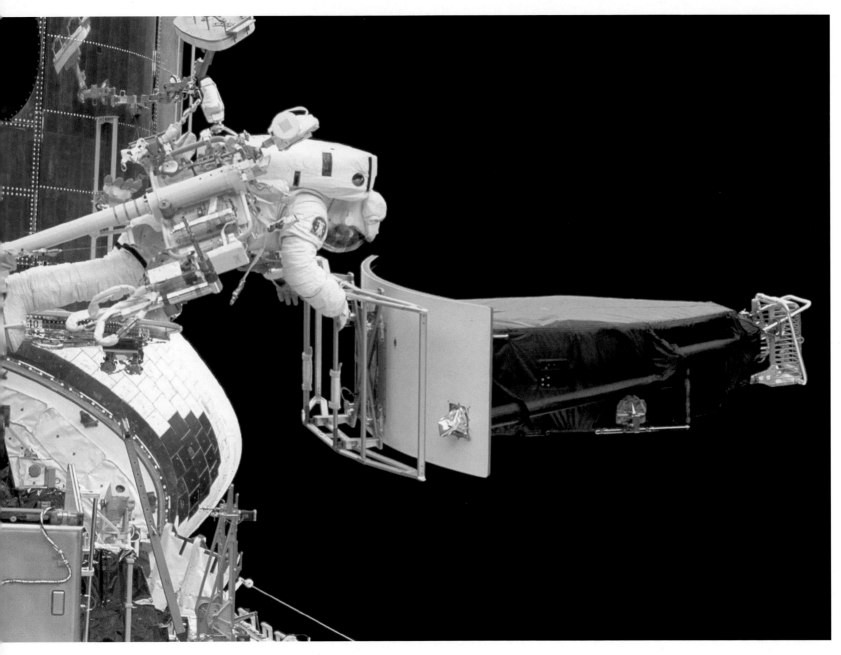

LEFT AND RIGHT: Astronauts working on the HST during the repair mission in which an upgraded Wide Field and Planetary Camera (known as WFPC2) was fitted. Although on earth it was the size of a grand piano and weighed 610lb, the astronauts could move it with ease in the zero gravity of an Earth orbit. This camera has subsequently produced many of Hubble's most dramatic images.

LEFT: The Hubble Space Telescope is
backdropped against black space as the
Space Shuttle *Columbia*, with a crew of seven
astronauts on board, eases closer March 3,
2002, in order to latch its 50ft-long robotic
arm onto a fixture on the giant telescope.

RIGHT: Space telescope orbiting the earth, September 1, 1999.

RIGHT: The final-polished 94in primary mirror being inspected prior to application of the reflective coating. During final shaping and polishing the approximately one-ton primary mirror was reduced to a weight of about 1,825lb.

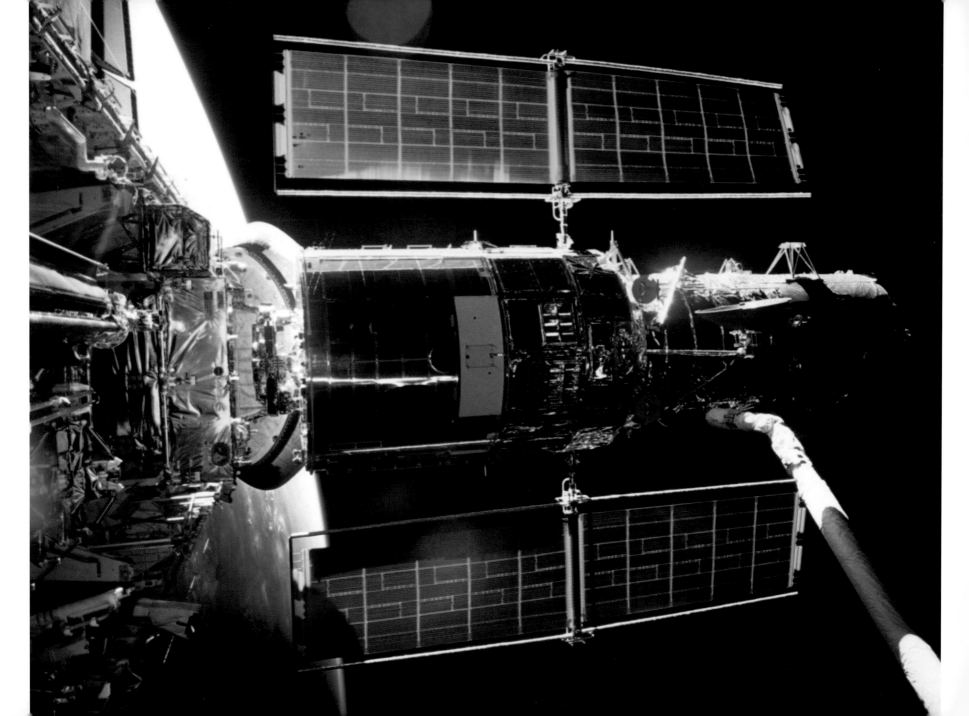

LEFT AND RIGHT: The Hubble Space
Telescope is retrieved by the Space Shuttle
Endeavour for repairs December 2–13, 1993.

LEFT: NASA astronauts replace the optics and gyroscopes on the flawed HST during mission STS-61.

RIGHT: The HST drifts away from the Space Shuttle *Endeavour* after its optics, gyroscopes, and solar panels were repaired.

PAGES 90/91: The Sun and Earth seen from the Space Shuttle. A NASA photograph.

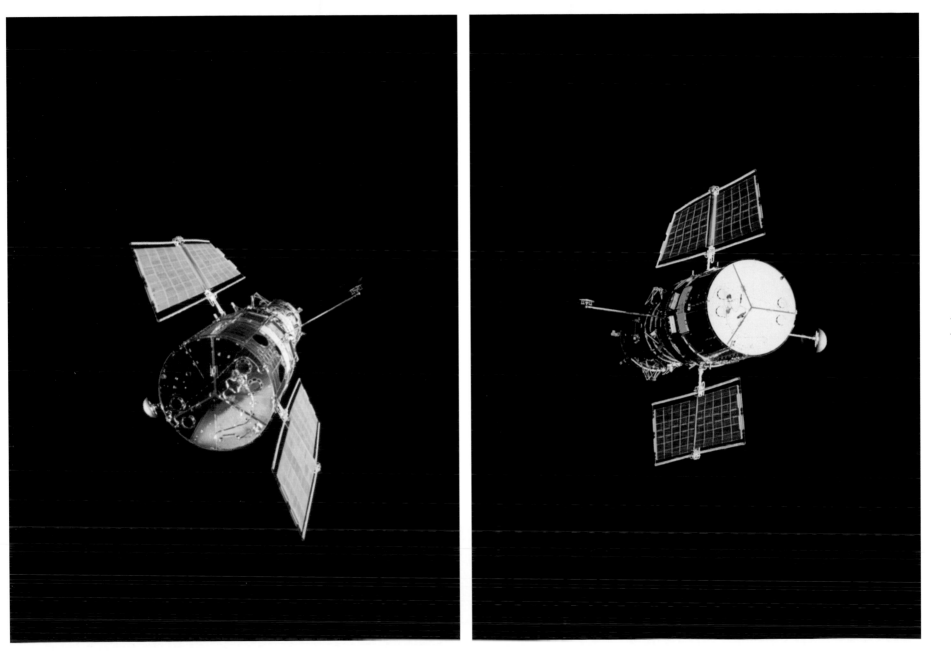

The Solar System

The planet Earth is one of eight which orbit around a central star, the Sun. The whole group is known as the Solar System and the pattern must be repeated countless times throughout the universe. In our own system, Earth is the third planet in terms of distance from the Sun, with Mercury the closest and Venus next. Outside earth are Mars, Jupiter, Saturn, Neptune, and Pluto. Earth has a single moon, the most conspicuous object in the night sky, while Mars has two very small satellites. However Jupiter has no fewer than 12 moons, of which four can be clearly seen using only binoculars or a low powered telescope. Saturn has ten moons, one of which, Titan, is the largest in the solar system and is even larger than the planet Mercury. Uranus has five satellites and Neptune only two. Pluto is generally regarded as the outermost planet although its eccentric orbit actually brings it closer than Neptune at times. Both Neptune and Pluto were discovered (in 1846 and 1930 respectively) as a result of mathematical computations based on Newton's work.

The planets are not the only bodies in the solar system. Between Mars and Jupiter are a group of relatively small objects known as the Asteroids. The largest of these is Ceres with a diameter of 480 miles, but there are thousands of smaller objects and various theories have been put forward to explain their existence. One is that they are the results of the collision and break up of two planets in the distant past, while another postulates that they are the remains of the original solar nebula which eventually formed the other planets. Some asteroids have pronounced eccentric orbits which can bring them close to earth. In 1937 one named Hermes came within 500,000 miles—only twice the distance to the moon and insignificant in astronomic terms. Today modern technology allows the movement of all asteroids to be closely monitored and contingency plans are being drawn up to cover the possibility of one on a collision course with Earth.

Another spectacular category of visitor to the solar system is the Comet. Basically they are lumps of ice attracted by the sun's gravitational field into massive elliptical orbits which bring them into view at various predictable times. As they come near to the sun, the outer layer vapourizes and forms a coma surrounded by hydrogen gas. Solar radiation and solar wind force the gases away from the direction of the sun, forming a bright tail which is often

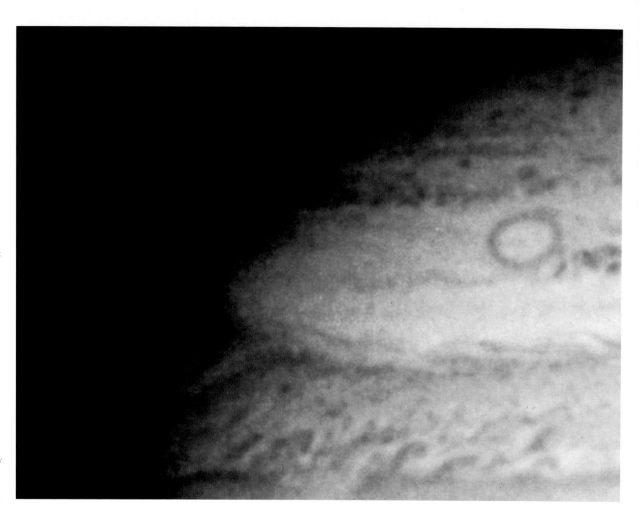

visible to the naked eye. Perhaps the most famous is Hailey's Comet which appears every 75 years, the last occasion being in 1986.

Designed to look into deep space, the Hubble telescope is not often used to look at objects within the solar system, but nevertheless is capable of producing some dramatic images. However, similar and more closeup images have been obtained from a variety of spacecraft such as the Voyager series which have flown close to the various planets, and other craft which have actually landed to send back images from or near the surface.

ABOVE: Although launched in 1990, it was not until March 11, 1991, that the Hubble was first directed at Jupiter. This view shows a closeup of the southeast quadrant of the planet a striking oval-shaped dark ring can be observed. This, and the famous Red Spot just out of view, are thought to be hurricane-like systems which carry ammonia ice crystals up to the upper levels of the atmosphere. Over the ensuing years, the Hubble has made regular observations so that these atmospheric systems can be tracked and monitored.

RIGHT: A closeup of Jupiter's surface showing the crater left by the impact of a fragment of Comet Shoemaker-Levy 9 which hit the planet on July 18, 1994. The smaller mark on the left was the result of the impact of a smaller fragment on the previous day. To give some idea of scale, the dark crescent shaped shadow has a diameter of 7,460 miles while the inner ring, possibly a shock wave from the impact, has a radius of 2,330 miles when this image was taken only 90 minutes after the impact.

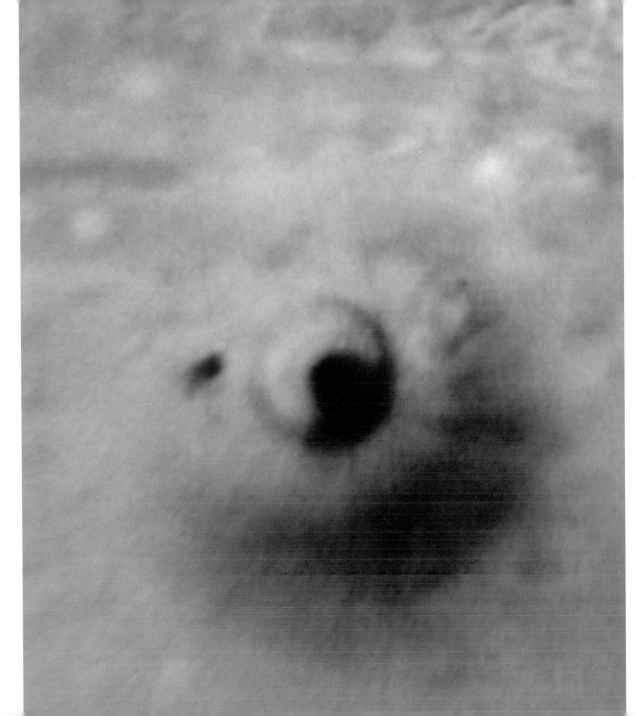

LEFT: Jupiter is the largest planet in the solar system with a diameter of 89,000 miles compared to Earth's 8,000 miles. In fact, in terms of mass, the combination of the sun and Jupiter account for 99.9 percent of the material in the whole solar system! This first true color photo of the mighty planet was taken in 1991 using the Wide Field Planetary Camera on the Hubble Telescope and shows the belts of cloud made up of frozen crystals containing many basic chemicals and elements including Ammonia, Carbon, Sulphur, and Phosphorus at a temperature of around -155°C (-247°F). Jupiter has several satellites or moons and the four largest of these can easily be seen with the aid of a low magnification telescope or binoculars.

FAR LEFT: Of all the objects in the solar system it is Mars, the Red Planet, which seems to fascinate us most. Perhaps it is the thought that it is relatively near (only 53 million miles when the shot was taken in December 1990) and that one day in the not too distant future it might be possible for man to actually travel there. The photo shows a layer of cloud of the north polar regions, but otherwise the atmosphere is clear and several features can be seen including a large dark area known as the Syrtis Major Planitia and which was first identified by astronomers as far back as the seventeenth century.

RIGHT: Saturn is the most beautiful of the planets to observe, adorned as it is by the spectacular multicolored rings which have a maximum diameter of 171,000 miles. By contrast they are exceedingly thin, latest measurements indicating a depth of only around 30ft so that when viewed edge on, they virtually disappear. The rings are actually made up of dust particles trapped in chunks of ice and the unique shape is formed by the effects of the gravitational effects of the huge planet and its ten satellite moons. These include Titan which is the largest moon in the solar system and is larger than the planet Mercury.

LEFT: The solar system is occasionally host to visitors from the deeper regions of space in the form of comets and meteorites. The passage of such bodies is usually the cue for great scientific activity. Of particular interest was Comet 9P/Tempel 1 which passed relatively near Earth (83 million miles) in the summer of 2005. This image was taken on June 30 and shows a dusty coma around the nucleus, although the Hubble telescope was not able to resolve the latter at this distance. However a closer view was to become available a few days later when the Deep Impact spacecraft was deliberately set in collision with the comet in order to release material and debris from the core which could then be analyzed by other spacecraft cameras.

RIGHT: Designed for looking out into deep space, the Hubble is rarely directed at some of the closer objects in the solar system. It cannot, for instance, point directly at the sun as this would damage the delicate camera sensors and the planet Mercury is also too close to be viewed. One way of gathering data from the sun is to analyze its reflected light and it was during such a task that this closeup of the 58-mile wide Copernicus crater on the moon was photographed. Such is the concentration of the Hubble's field of view that it would need to take 130 pictures to cover the whole of the moon's surface.

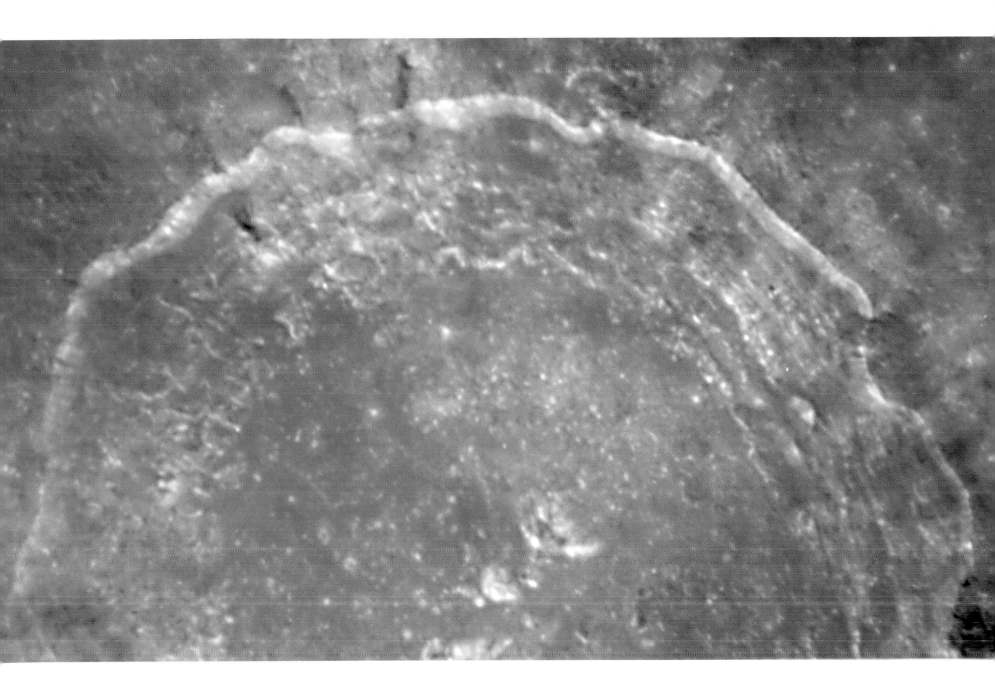

ABOVE: Ultraviolet image of Jupiter taken by the Wide Field Camera of the HST. The image shows Jupiter's atmosphere at a wavelength of 2,550 Angstroms after many impacts by fragments of comet Shoemaker-Levy 9.

ABOVE: The Hubble telescope was used to keep an eye on changes in Jupiter's turbulent atmosphere while planning for the Galileo space probe's arrival at the giant planet. This image provides a detailed look at a unique cluster of three white, oval-shaped storms that lie southwest (below and to the left) of Jupiter's Great Red Spot (dark oval-shaped object near the picture's right edge). The appearance of the clouds is considerably different from their appearance only seven months earlier. Hubble shows these features moving closer together as the Great Red Spot is carried westward by the prevailing winds, while the white ovals are swept eastward.

LEFT: Five spots—one colored white, one blue, and three black—are seen scattered across the upper half of Jupiter. They are actually a rare alignment of three of Jupiter's largest moons: Io, Ganymede, and Callisto. In this image, the telltale signatures of this alignment are the shadows (the three black circles) cast by the moons. Io's shadow is located just above center and to the left; Ganymede's on the planet's left edge; and Callisto's near the right edge. Only two of the moons, however, are visible in this image. Io is the white circle in the center of the image, and Ganymede is the blue circle at upper right. Callisto is out of the image and to the right.

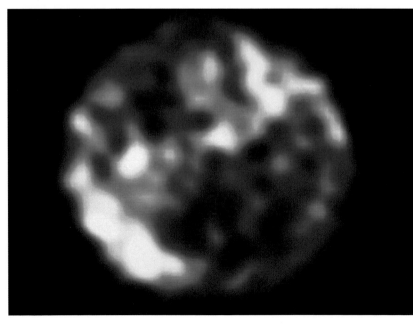

ABOVE: In a dress rehearsal for the rendezvous between NASA's Deep Impact spacecraft and comet 9P/Tempel 1, the HST captured dramatic images of a new jet of dust and gas streaming from the icy comet. The images are a reminder that Tempel 1's icy nucleus, roughly half the size of Manhattan, is dynamic and volatile. The eruption of dust seen in these observations was a preview of the fireworks that came in early July 2005, when a probe from the Deep Impact spacecraft slammed into the comet.

RIGHT: Detail of Jupiter's moon Io—the best images of this since a pair of Voyager spacecraft flew by the small moon some years ago.

ABOVE LEFT: Frosty white water ice clouds and swirling orange dust storms above a vivid rusty landscape reveal Mars as a dynamic planet in this sharp HST view, when Mars was a close—approximately 43 million miles—from Earth. Especially striking is the large amount of seasonal dust storm activity seen in this image. One large storm system is churning high above the northern polar cap (top of image), and a smaller dust storm cloud can be seen nearby. Another large duststorm is spilling out of the giant Hellas impact basin in the Southern Hemisphere (lower right).

ABOVE: Springtime on Neptune—a 2002 image.

ABOVE LEFT: This color image of Saturn was taken with the HST's Wide Field and Planetary Camera (WF/PC) in the wide field mode on August 26, 1990, when the planet was at a distance of 860 million miles from Earth.

ABOVE: The dancing light of the auroras on Saturn behaves in ways different from how scientists have thought possible for the last 25 years. New research by a team of astronomers led by John Clarke of Boston University has overturned theories about how Saturn's magnetic field behaves and how its auroras are generated.

LEFT: This image is from Hubble's infrared camera, which has taken its first peek at Saturn and provides detailed information on the clouds and hazes in Saturn's atmosphere. The blue colors indicate a clear atmosphere down to the main cloud layer. Most of the Northern Hemisphere that is visible above the rings is relatively clear. The dark region around the South Pole indicates a big hole in the main cloud layer. The green and yellow colors indicate a haze above the main cloud layer. The red and orange colors indicate clouds reaching up high into the atmosphere. The rings, made up of chunks of ice, are as white as images taken in visible light.

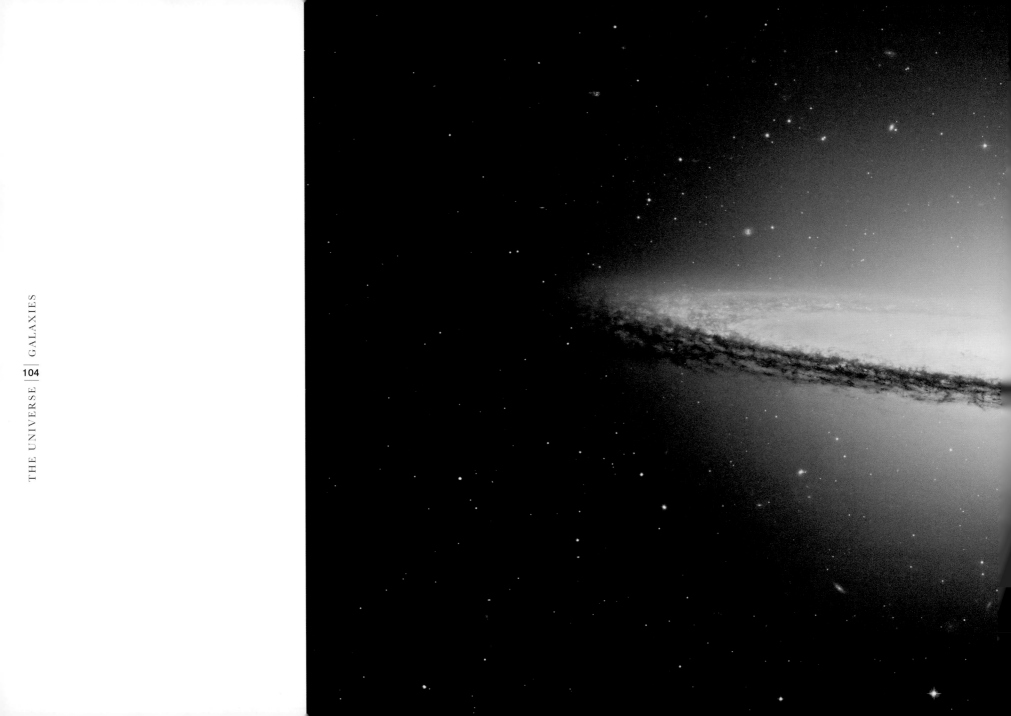

LEFT: Another spectacular image from the distant reaches of the universe. The so-called Sombrero Galaxy is officially known as Messier 104 (M104) and is a typical spiral galaxy viewed edge on. Located in the constellation of Virgo, an area rich in galactic clusters, it is 28 million light-years from Earth and is calculated to span no less than 50,000 light-years. Another fascinating statistic was provided by the astronomer V. Slipher who calculated in 1912 that the whole galaxy was moving away from us at the fantastic speed of 700 miles per second. This supported the theory of a constantly expanding universe.

Galaxies

A galaxy is a discrete concentration of numerous stars, separated from similar formations huge distances. A single galaxy can contain literally billions of stars and can stretch across thousands of light-years in space. Our own sun is an insignificant part of the Milky Way, a galaxy which is thought to contain over 200,000,000,000 stars in a disc-like pattern some 3,000 light-years thick and having a diameter of over 100,000 light-years. Of spiral shape, it has been calculated that the Milky War rotates at the rate of one revolution every 200 million years. And the Milky Way is itself the merest fragment of the whole universe which consists of literally millions of other galaxies. Other nearby galaxies include the Large and Small Magallenic clouds visible from the southern hemisphere—relatively close at 170,000 and 200,000 light-years respectively. Yet these are classified as dwarf galaxies, virtually satellite systems to our own Milky Way. The nearest large galaxy is the Andromeda Galaxy (M31) which is 2.4 million light-years away and is just visible to the naked eye as a faint patch. Viewed through a telescope, its spiral shape is apparent and it also has its attendant dwarf galaxies.

While spiral galaxies are quite common, there are many other shapes including elliptical galaxies which range from elongated ovals to almost spherical. Other galaxies have no clearly defined shape. Within the galaxies are other elements which can often be observed directly through telescopes or through various filters (such as infra red) or by the use of radio astronomy. This include clusters of stars known as Open or Globular. The former consist of up to 500 newly formed stars in a relatively small area and are generally close to the plane of the galaxy. Some are still shrouded in gas, indicating that star formation in ongoing while others such as the Pleiades and Hydra, are much older and stable. Globular clusters contain from 50,000 to a million stars and in the case of the Milky Way are not confined to the plane of the galaxy but tend to orbit the center of the galaxy at distances of 50-60,000 light-years. They generally consist of stars some 12,000,000,000 years old, going back to the time that the galaxy was first evolved.

Apart from the Milky Way and the Andromeda Galaxy, there are some 20 galaxies grouped together as a cluster. They move about a common center of mass located between the Andromeda and the Milky Way. It is estimated that the total mass of all the stars and other bodies in this group is equivalent to 500 billion suns. Our own local group is not unique. There are similar groups of galaxies throughout the universe. In the area north of the star Spica, in the constellation Virgo, is one of the largest containing tens of thousands of galaxies centered some 40 million light-years from the solar system.

ABOVE: A fascinating view of dust clouds formed in the elliptical Galaxy NGC1316, part of a cluster of galaxies on the edge of a Fornax, a southern constellation. This formation is around 75 million light-years away and is also the source of some of the strongest and largest radio signals in the universe. Measurement of red star clusters in the galaxy have led scientists to suggest that the present formation is the result of the merger of two separate spiral galaxies, an event which occurred billions of years ago. Detailed examination of the area has revealed a variety of patterns in the swirls of dust including ripples, loops, and plumes which are light-heartedly referred to as dust bunnies.

RIGHT: One of the most distinctive galaxies, the Whirlpool Galaxy is officially designated M51 (NGC5194). This view taken in 2005 clearly illustrates the graceful form of this group of stars, connected by swirling clouds of dust and gases. The galaxy is located some 31 million light-years away and is located in the constellation Canes Venatici (Hunting Dogs) faintly visible in the northern sky below the handle of the Big Dipper, or Plough. The smaller constellation, which appears connected to one of the Whirlpool's flailing arms, is actually NGC5195. It is, in fact, well behind the larger galaxy.

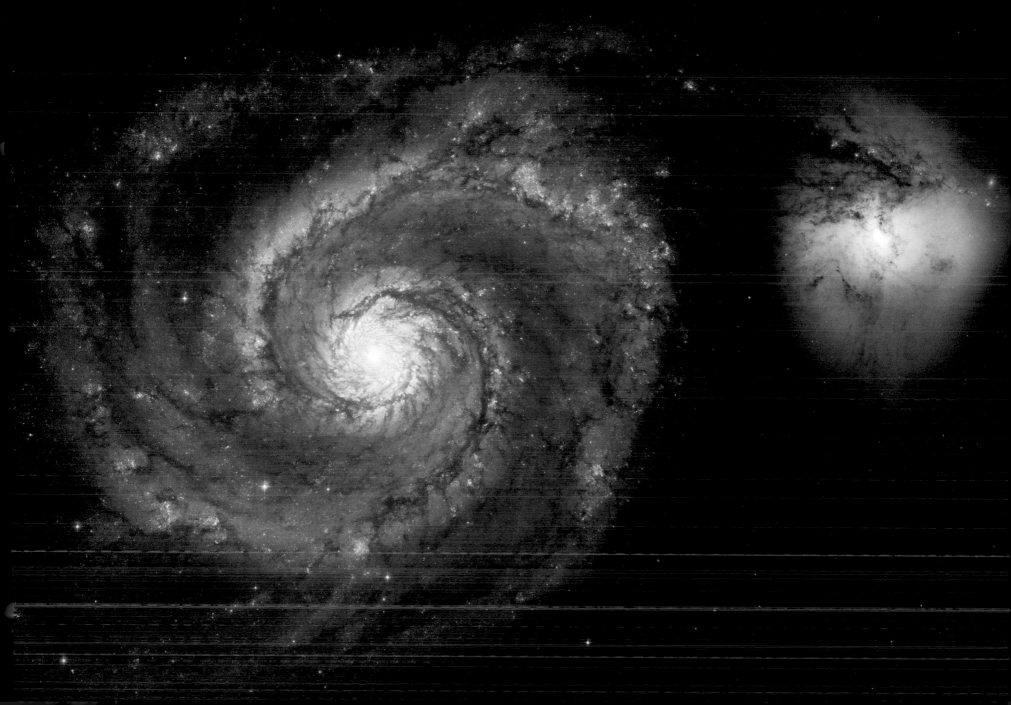

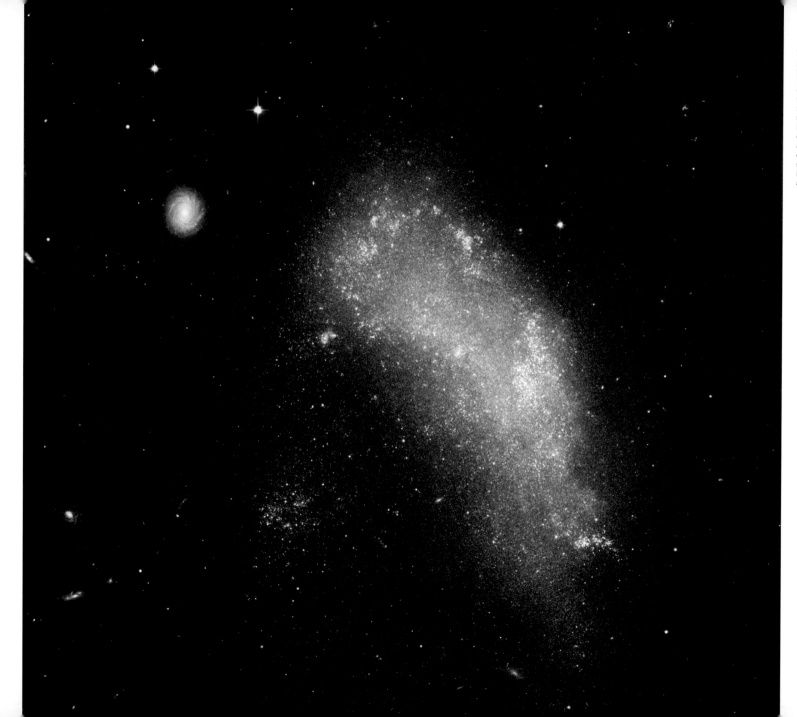

LEFT: This 2003 image shows the irregular galaxy NGC1427A, some 62 million light-years away near the constellation Fornax. Under the gravitational force of the cluster of larger galaxies in this vicinity the gas within NGC1427A is collapsing and forming new stars, many of which are visible in this shot. Eventually these stars will be drawn into the other galaxies and this cluster will cease to exist as an identifiable mass, although the whole process will take billions of years. In the top left background can be seen a more established spiral galaxy but this is over 1,500 million light-years away—an almost unimaginable distance.

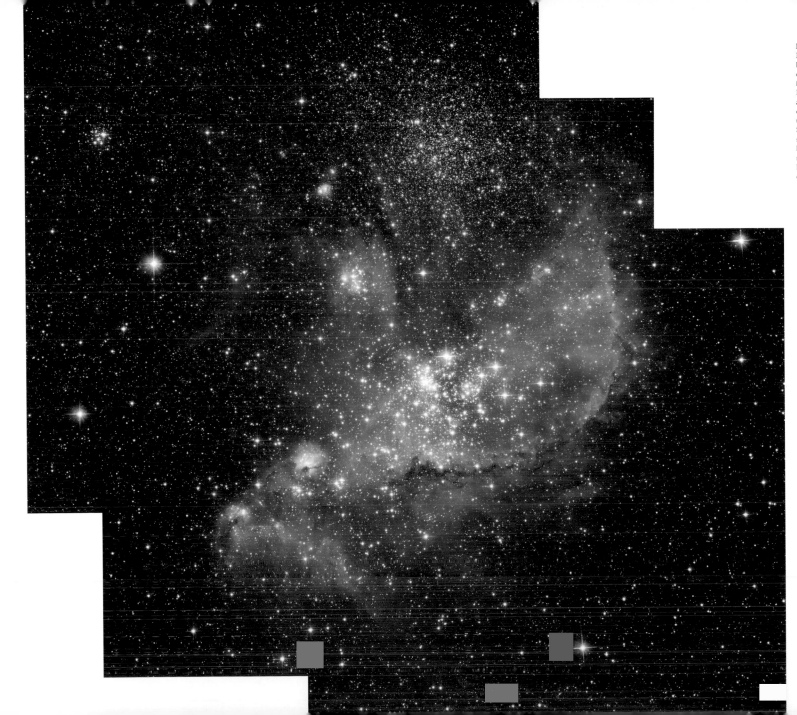

LEFT: This object, known as the Small Magellanic Cloud, is located in the southern hemisphere constellation Tucana on the edge of the Milky Way. It is only 210,000 light-years away. Otherwise known as nebula NGC346, it exhibits numerous stars which are in the early stages of formation from collapsing gas clouds which have not ignited their hydrogen to sustain nuclear fusion. Scientists believe that fragmentary galaxies such as this form the building blocks of larger galaxies and it has been extensively studied to improve our understanding of how stars are created. This nebula alone contains over 2,500 infant stars.

LEFT AND (DETAIL) RIGHT: In this breathtaking view, a ring of brilliant blue stars surrounds the yellowish mass of what was once a normal spiral galaxy. This galaxy, rather unromantically designated AM0644-741, is over 300 million light-years away in the direction of the southern constellation Volans. The whole measures some 150,000 light-years across making it larger than our own Milky Way. The formation of this galaxy resulted from the collision of two galaxies and can be likened to a stone being thrown into a pond forming a ring of ripples, but in this case on the grandest scale imaginable.

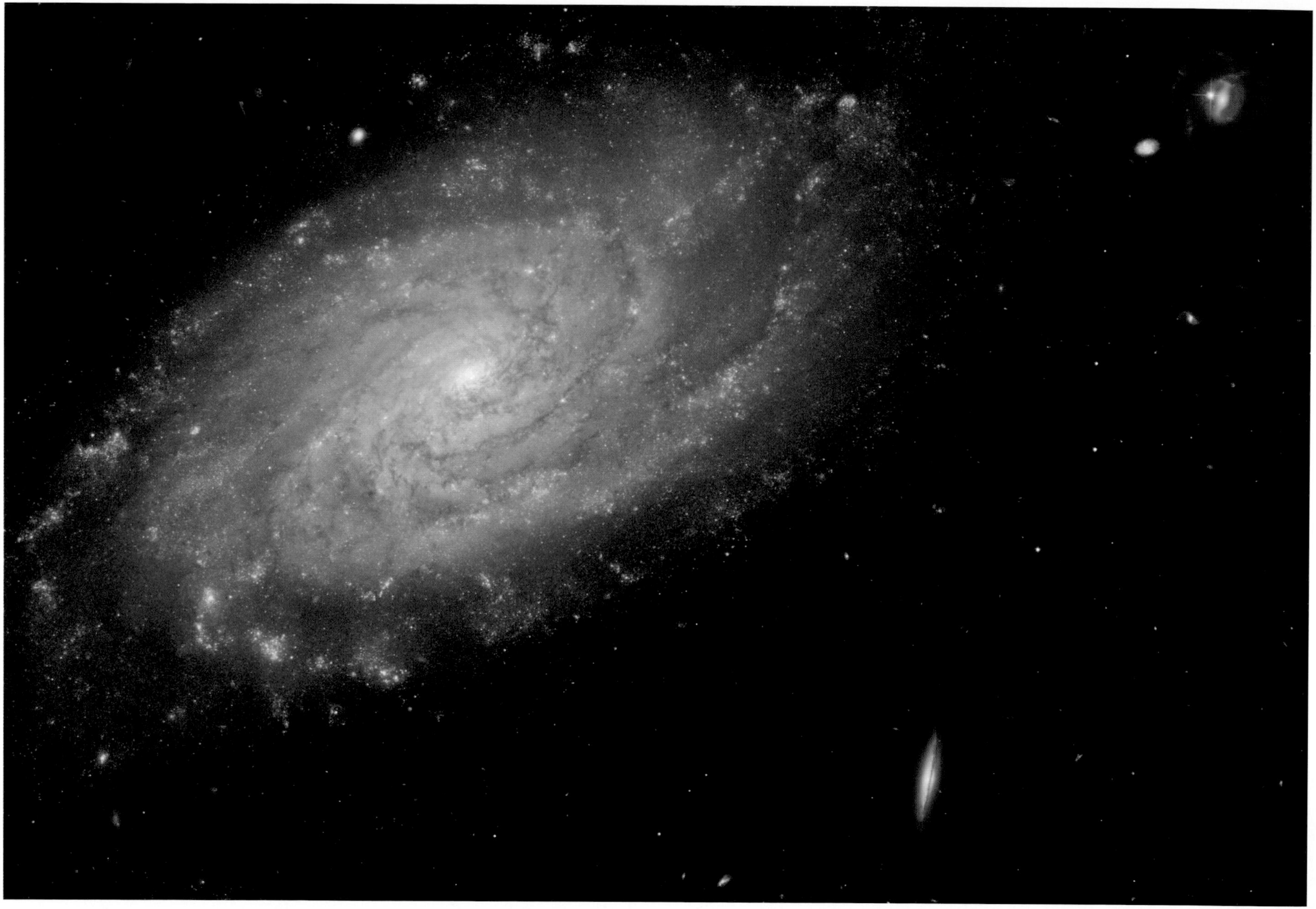

LEFT: Inevitably this unusual galaxy (UGC10214) has been dubbed the Tadpole due to its distinctive shape. It is located some 420 million light-years away in the constellation of Draco. Its long tail is the result of a collision with a smaller galaxy, just visible as a small spot of light in the top left-hand corner of the tadpole's head, and is virtually a trail of debris although it is a staggering 280,000 light-years long. The two brighter patches of stars will, in time, form small separate galaxies. The clusters in the head of the tadpole are tinged with blue, indicating that they contain several massive stars 10 times hotter and one million times brighter than the sun.

FAR LEFT: Returning to the northern hemisphere, this dusty spiral is known as NGC3370 and resides in the constellation of Leo. Almost 100 light-years away from Earth, this galaxy has proved important to astronomers in a number of respects. In 1994 an exploding star or supernova (SN1994ae) within the galaxy was observed in great detail and provided considerable data on the growth rate of the expanding universe. In addition other fainter stars of known varying brightness—known as Cepheids—allow accurate measurements of distance to be taken. This work has been vastly boosted by the capabilities of the Hubble telescope which is able to resolve the individual Cepheids.

LEFT: The faint patch in the background of this image is the Sagittarius Dwarf Irregular Galaxy, more conveniently known as the SagDIG. The bright stars with spiky glare effects are actually nearby stars in our own Milky Way but the fainter bluish stars of SagDIG are some 3.5 million light-years away. It is of interest to scientists as it shows many stars in the early stages of development in a gas-rich galaxy so that it can be used to validate theories concerning the influences which trigger star formation. It is only the advent of the Hubble telescope with its ability to resolve individual stars at this great distance which has made this work possible.

FAR LEFT: Looking through an ordinary telescope, most stars and celestial objects appear white, with perhaps a tinge of blue, red or yellow. However, the ability of the Hubble telescope to view different light wavelengths can produce startlingly different images. A case in point is this ultraviolet view of the Galaxy NGC6782 stationed 183 million light-years from Earth. In the normal visible light spectrum it appears as a conventional tightly wound spiral, but viewing the ultraviolet light shows up the hotter stars and the bright central ring shows that the galaxy contains many recently formed very hot stars. Retreating galaxies further away are subject to "red shift" where their light spectrum is stretched to the longer wavelengths. In order to compare the brightness of galaxies at varying distances, the ability to view nearby galaxies in the shorter ultraviolet wavelength is, therefore, important.

LEFT: Shown here is Galaxy NGC300 which is a spiral galaxy similar to the Milky Way. It is also a near neighbor that it is only some 6.5 million light-years away from Earth. It is one of group of galaxies known as the Sculptor group in the southern constellation of the same name. There are so many stars that they appear almost as grains of sand but in fact they include a number of young massive stars known as Blue Supergiants which are among the brightest stars in the sky but their great distance makes them only now visible to the lens of the Hubble telescope.

117

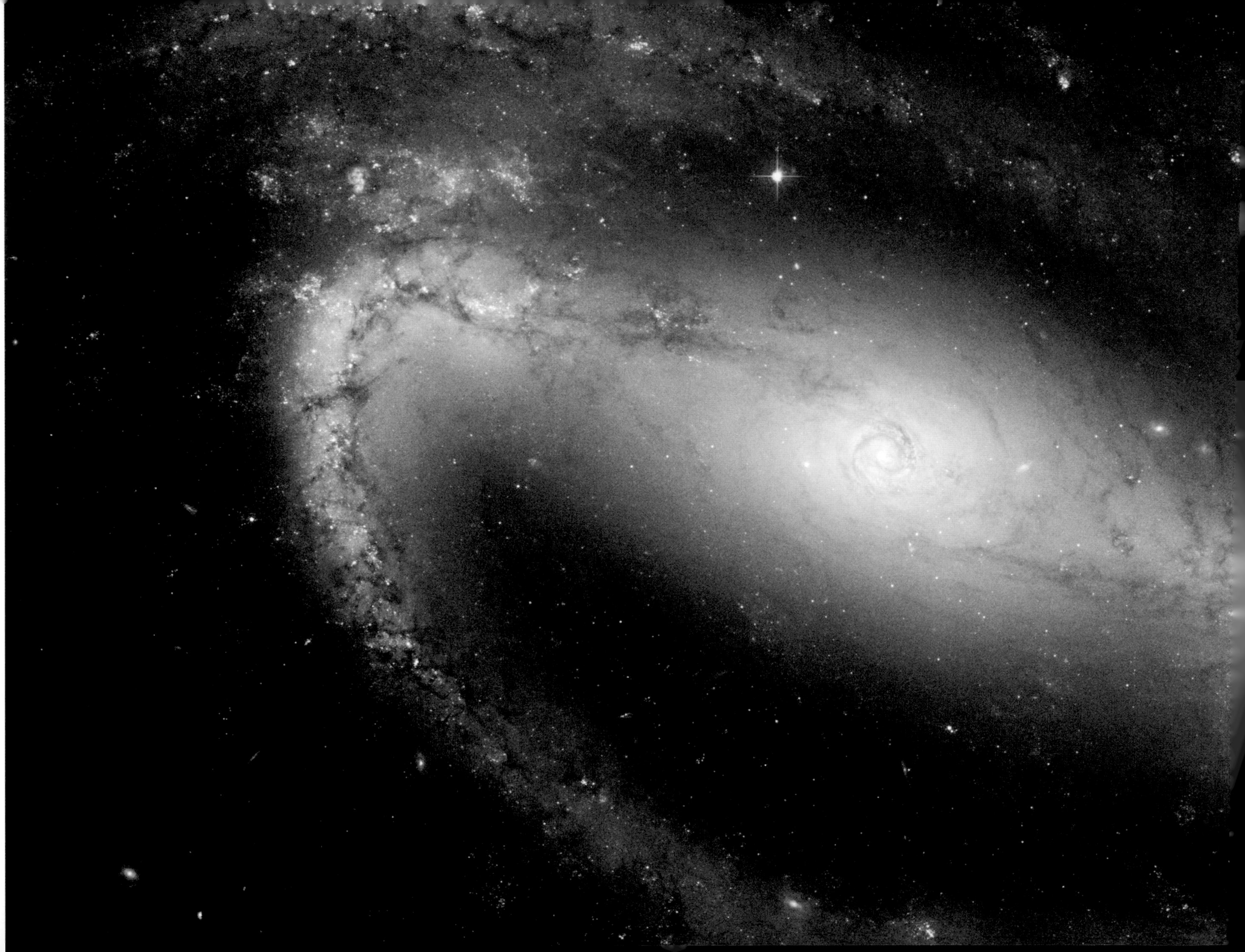

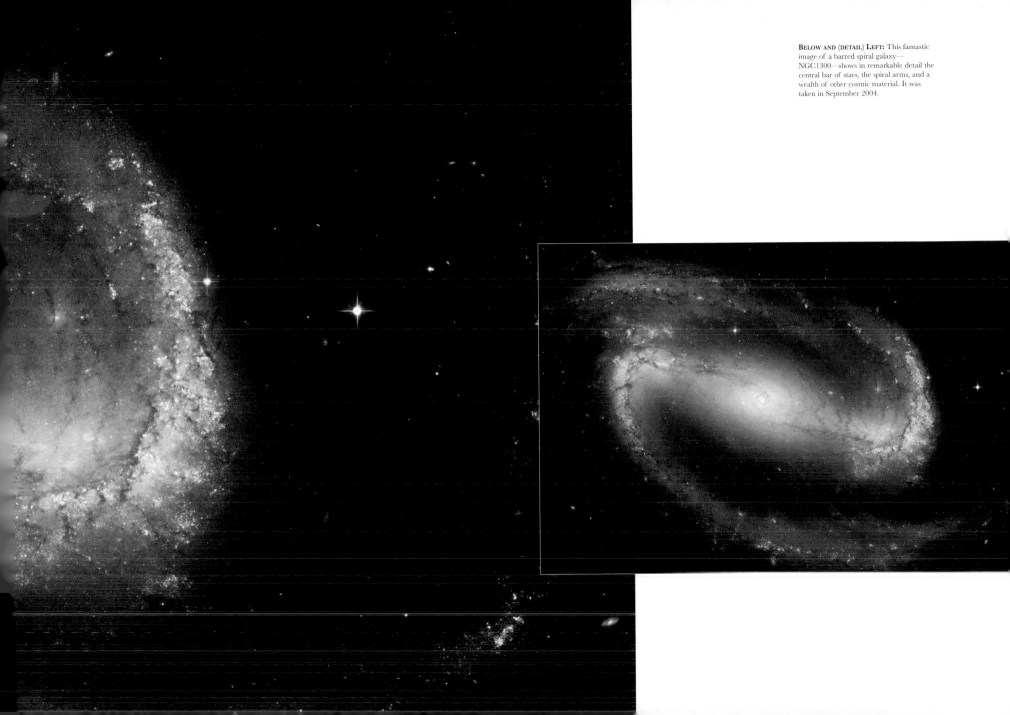

BELOW AND (DETAIL) LEFT: This fantastic image of a barred spiral galaxy— NGC1300—shows in remarkable detail the central bar of stars, the spiral arms, and a wealth of other cosmic material. It was taken in September 2004.

LEFT: This view of a distant part of the universe shows remote clusters of galaxies (CL0939 and 4713). Because of their great distance, we are actually looking at how this part of the universe looked when it was two thirds of its present age and so the Hubble telescope enables astronomers to look at ancient galaxies. Almost all are variations of the basic spiral formation but there are "fragments" of other galaxies interspersed. This image was obtained during January 1994 using the Wide Field Camera 2.

RIGHT: A galaxy collision in NGC6745. A large spiral galaxy, with its nucleus still intact, peers at the smaller passing galaxy (nearly out of the field of view at lower right), while a bright blue beak and bright whitish-blue top feathers show the distinct path taken during the smaller galaxy's journey. These galaxies did not merely interact gravitationally as they passed one another, they actually collided.

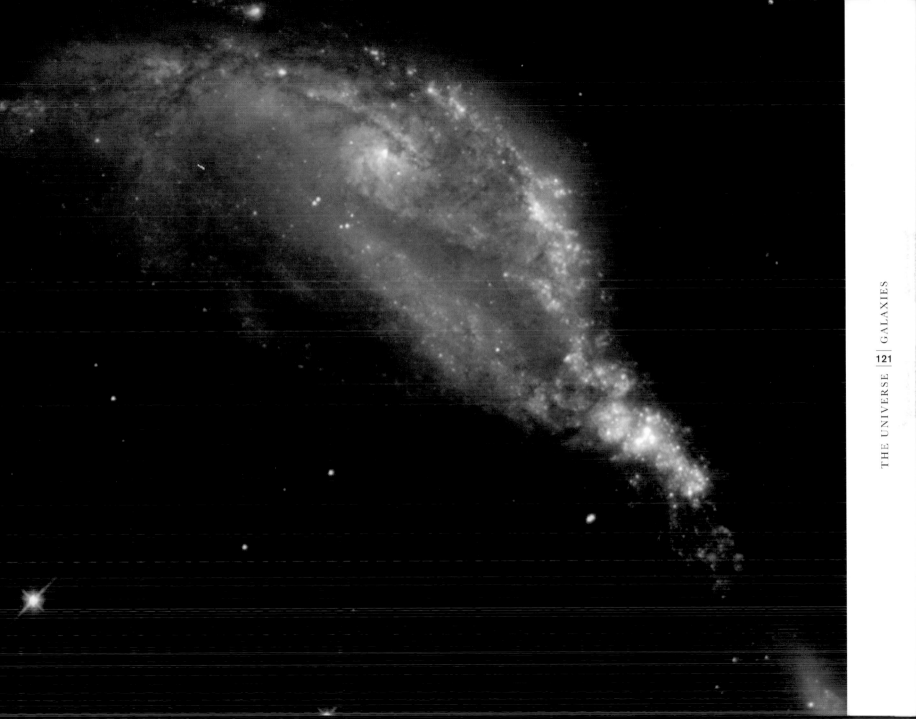

LEFT: One of the classic spiral galaxies is the aptly named Whirlpool Galaxy in the constellation of Canis Venatici, part of the northern sky adjacent to the well-known Plough formation. Otherwise known either as M51 or NGC5194, the Whirlpool is one of the most popular objects in the sky for amateur and professional astronomers as it is easily viewed and even photographed with the aid of relatively low-powered telescopes. However, the technology available to Hubble scientists enables them to more clearly view the dust clouds and banks of hot hydrogen gases and associate them with individual star clusters within the galaxy.

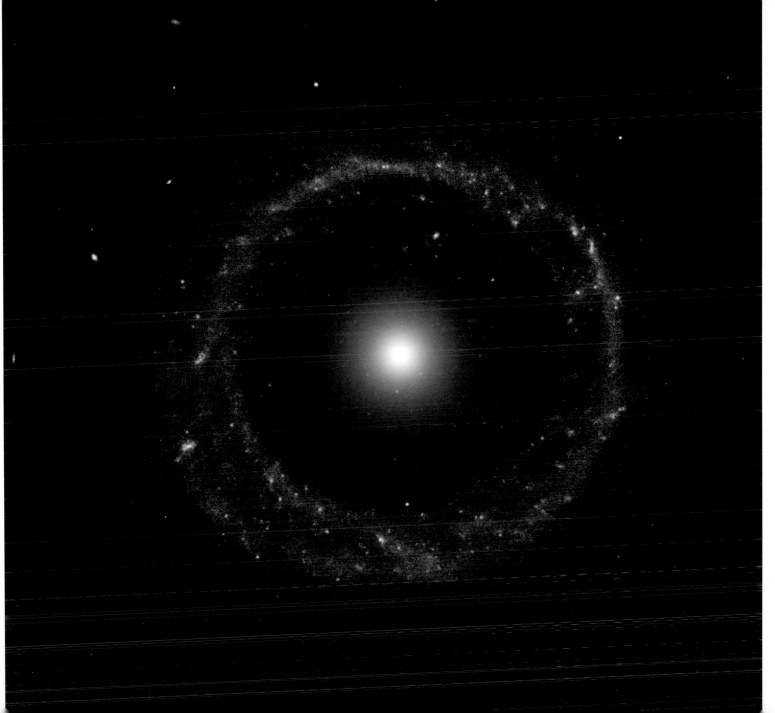

LEFT: This unusual galaxy with its distinctive "smoke ring" was discovered in 1950 by astronomer Art Hoag and is now universally known as Hoag's Object. The entire galaxy is some 120,000 light-years wide, slightly larger than the Milky Way. The distinctive outer ring is formed of hot blue stars which surround a nucleus of colder yellow stars. How this configuration came about is a matter of conjecture although some scientists favor the theory that it was the result of a collision between two galaxies some two to three billion years ago. Hoag's Object lies some 600 million light-years away in the equatorial constellation of Serpens.

RIGHT: Many stellar objects have evolved their current form as the result of a collision with other objects some time in the distant past. However, in this case we can see such an event occurring. These two galaxies (NGC2207 on the left, IC2163 on the right) in the constellation of Canis Major are in the process of a major interaction. However we will not live to see the outcome as it is calculated that the closest point of contact occurred some 40 million years ago

LEFT: The bright area in this picture is Galaxy NGC1275 embedded in the center of a cluster of galaxies known as the Presses Cluster situated some 235 million light-years away. It is known as a source of strong radio and X-ray emissions, possibly indicating the existence of a black hole at the center of the galaxy. The dark streaks across the bright background are traces of another spiral galaxy viewed edge on and the two are approaching each other at a speed of over 6 million miles per hour. As an example of the sophistication of the Hubble imagery, this picture was made of archive red and blue images taken in 1995 combined WFPC2 infra red observations in 2001.

LEFT: A closeup of the core of the Whirlpool Galaxy showing a marked X-shape silhouetted against the bright nucleus. The shadow is due to dust absorption and marks the position of a black hole whose mass is equivalent to one million suns. The vertical bar is thought to be an edge-on dust ring 100 light-years in diameter which masks the black hole and the accretion process from direct observation. The cross arm of the X may be another dust ring.

LEFT: While distant objects are fascinating in their own right, exciting developments occur much closer to home. The nearest active galaxy to Earth is Centaurus Alpha, visible from the southern hemisphere. This closeup shows the extensive dust lane which surrounds the entire galaxy, believed to be the remnant of a smaller spiral galaxy, which merged with the much larger elliptical galaxy. The shock of the collision compressed the interstellar gas, triggering the formation of numerous new stars. In this dramatic image, the dark dust clouds are backlit by the glow of hot gases and the star fields behind it.

RIGHT: This group of tightly packed galaxies in the constellation Serpens is known as Seyfet's Sextet after the astronomer Carl Seyfert who discovered them in 1940. There are actually only five galaxies but they are so tightly packed that their gravitational interaction has caused a tail of stars resembling a galaxy to form off the southeast arm of the group. The whole group spans only 100,000 light-years which is less than the size of the Milky Way. In the course of billions of years the separate galaxies will interact, collide, and combine to form a single large elliptical galaxy.

PAGE 130: Another close grouping of galaxies and the first such to be discovered, in this case by Edouard Stephan in 1877 and therefore known as Stephan's Quintet. The intense gravitational forces generated by these huge formations in close proximity has been the trigger for the formation of millions of new stars. In this image there are three noteworthy areas of star formation which are the long sweeping tail and spiral arms of NGC7319 (near center), the gaseous debris of NGC7318A and 7318B (top right), and an area called the Northern Starburst Region (top left). The bluish tinge of these areas indicate that the stars are relatively young, only one or two billion years old.

PAGE 131: Those of us who think that 20 seconds at f5.6 is a long exposure for a camera will be amazed to learn that this Hubble image was made up of 350 separate exposures for a total exposure time of 3.5 days during the period December 2, 2002, to January 11, 2003. The subject is the halo of stars surrounding the Andromeda Galaxy (M31), the nearest spiral galaxy to earth but still some 2.5 million light-years away. The full image resolves some 300,000 stars which, unexpectedly, vary in age from 6 to 13 billion years old, in comparison to the more uniform 11–13 billion years of stars in the Milky Way

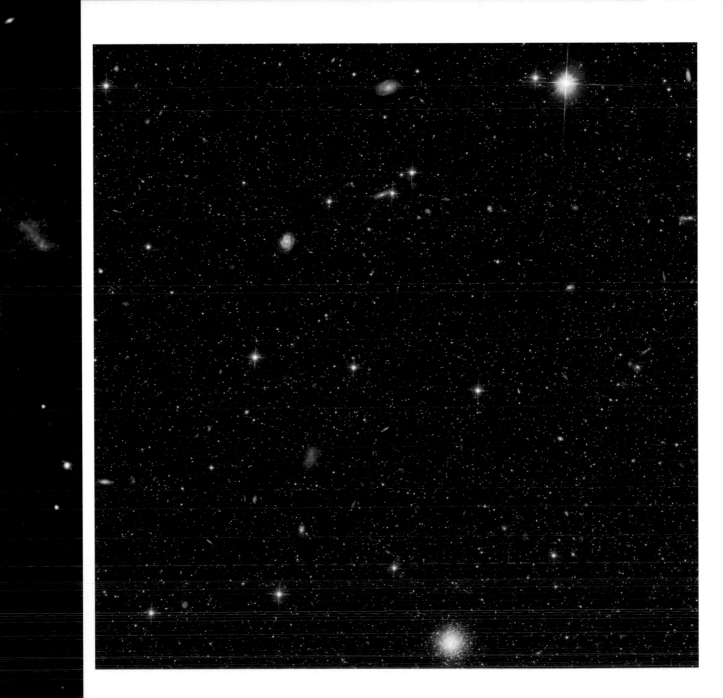

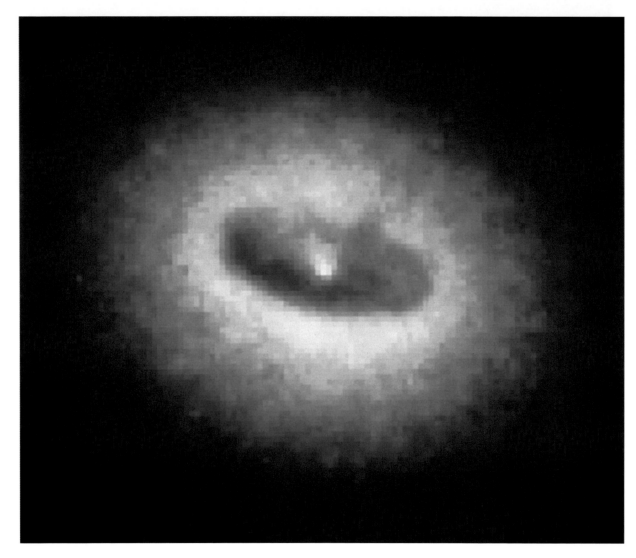

LEFT: The term "Black Hole" relates to the final stages of the life cycle of a large star when it collapses in on itself, forming a small object with such high mass that its gravitational force even prevents the escape of light energy. In this picture can be seen a giant disc of cold gas and dust fueling a possible black hole at the core of galaxy NGC4261. The disc is believed to be 300 light-years across. While its outer edges are cold, temperature increases toward the center where the ultrahot accretion disc is formed. Here intense gravitational forces compress and heat the material, the resulting excitation of the gases causing the emission of strong radio signals which provide astronomers with evidence of the black hole.

LEFT: The spiral galaxy NGC4603 in the Centaurus cluster some 108 million light-years away. The spiral arms are highlighted by clusters of young bright blue stars while the giant red stars in the process of dying can also be seen. The galaxy also contains many examples of a class of pulsating star known as a Cepheid Variable. The brightness of these stars constantly varies in a steady cycle which can last between one and fifty days. Such stars are very useful to astronomers in calculating the distance of galaxies in which they reside. Such measurements show that the enormous Centaurus group has a great effect on other galaxies such as our own Milky Way: it is calculated that we are being drawn toward Centaurus at a speed of one million miles per hour.

LEFT: Galaxy NGC3079 lies some 50 million light-years from Earth in the constellation Ursa Major. This closeup image of the galaxy's core shows a bubble of gaseous filaments rising 3,500 light-years above the plane of the galaxy's disc. Measurements show that these gaseous filaments are ascending at a staggering four million miles per hour. It is thought that this phenomenon arises from solar winds created by hot stars mixed with bubbles of very hot gas from supernova explosions. All in all, this indicates an energy release on a cataclysmic scale which will eventually result in the formation of numerous new stars.

RIGHT: This photograph, taken in 1991 and one of the earliest Hubble projects, shows the bright sphere which is the elliptical galaxy M87. The brightness at the center of the galaxy indicates that most of the stars are concentrated there by the gravitational field of a massive black hole. Computer models indicate that this could have a mass of 2.6 billion suns. The central spot of light is a visible manifestation of a nuclear emission caused by hot plasma interacting with magnetic fields which generates the very strong radio signals associated with black holes. The streak of light to the left is a plasma jet extending over 5,000 light-years from the nucleus.

LEFT: This spectacular pattern of colors is part of the Large Magellanic cloud, one of the nearest galaxies to the Milky Way and visible from the southern hemisphere. The starfields and wisps of gas provide a backdrop for Supernova 1987a—a massive star in the process of self-destruction. This complex three-colored image is made up of separate exposures taken on the Wide Field and Planetary Camera 2 in 1994, 1996, and 1997.

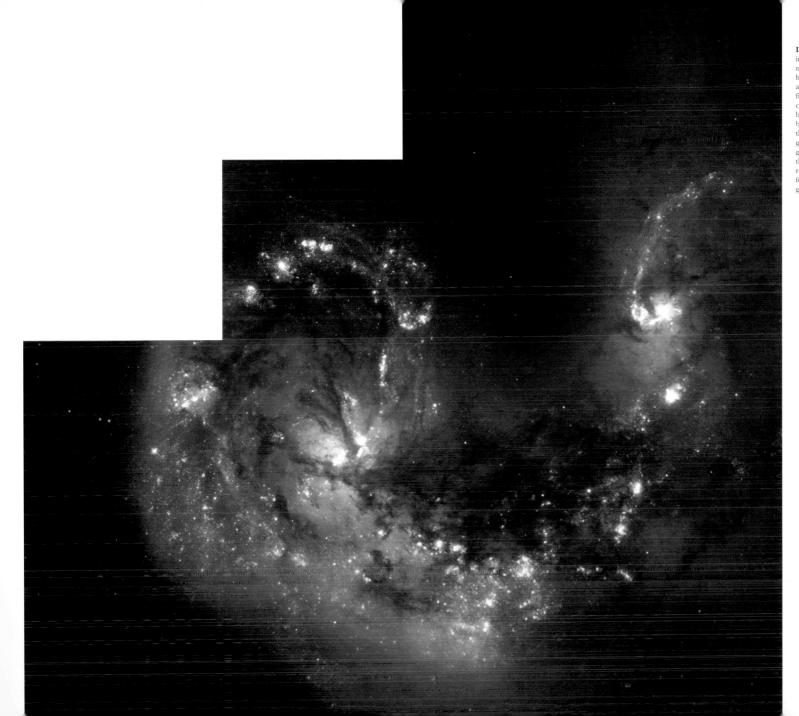

LEFT: Another example of galaxies colliding, in this case NGC4038 and 4039 situated 63 million light-years away in the southern hemisphere constellation Corvus. The cores are seen as orange blobs criss-crossed with filaments of dark dust, while further dust clouds stretch between the cores. The brighter spiral-like patterns are formed by bright clusters of newly formed blue stars, their creation being triggered by the forces generated by the collision of the two galaxies. Not visible in this closeup view are the two distinctive trails of stars which radiate outward and lead to the whole formation being known as the Antenna galaxies.

LEFT: As scientists became more practiced in the use of the Hubble telescope, and its optics and systems were upgraded, it became possible to look further and further into the remote corners of the universe. This photo shows a remote galaxy cluster known as Abell 2218I. In the left center of the picture, below and slightly to the right of the very bright mass on the left, are two very faint red images which are actually derived from the same object. This is believed to be a cluster of around one million stars, far fewer than a mature galaxy and therefore very young in astronomic terms. It is situated 13.4 billion light-years away from our solar system.

RIGHT: The bright object in this mosaic image taken in March 1994 is the 13th magnitude NGC4881 Galaxy on the edge of the Coma Cluster near the constellation of Virgo in the northern hemisphere. Apart from the smaller galaxy on the right and a few Milky Way bright stars, virtually everything else in this picture is much further away The Coma Cluster is 320 million light-years away and detailed observations of the brightness distribution of its globular star cluster are important in fixing its exact distance so that it can be used as a reference point when accurately measuring distances to more distant objects.

LEFT: At the center of this picture is the spiral Galaxy 0313-192 seen almost edge on. It is almost one billion light-years away from earth. What is particularly interesting is the manifestation of a massive radio-emitting jet along the axis of the spiral, the first time that such a phenomenon has been observed and captured by Hubble. The image was taken by the Advanced Camera for Surveys in July 2002.

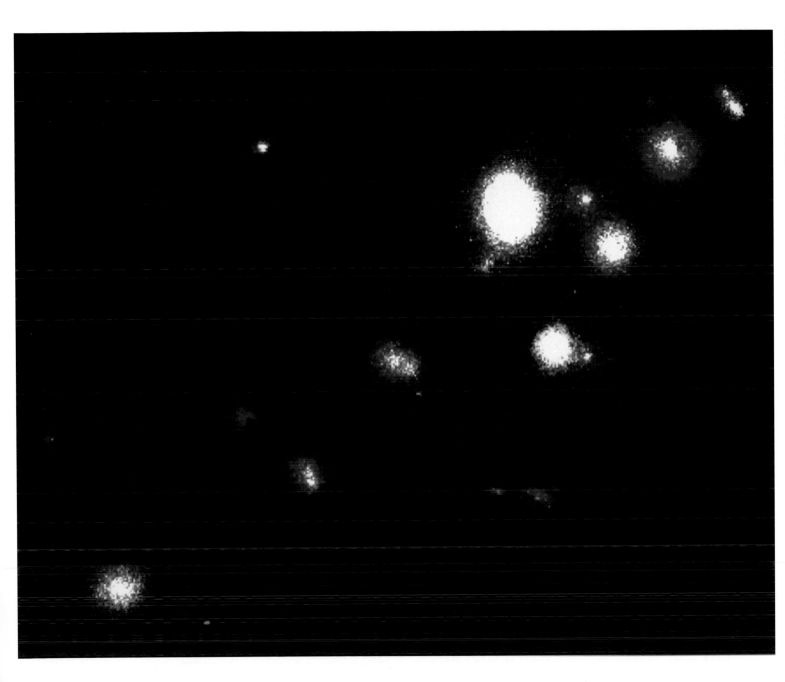

Although Hubble has produced some fascinating results, much of its work is involved in long-term surveys of huge areas of the night sky. Such areas will have been surveyed previously by earthbound telescopes and in many case on tantalizingly faint images were available or else nothing was seen at all. However, the greater clarity and resolution of the Hubble telescope means that these objects can be revisited and there shape and structure revealed in much greater detail. An example of this process is shown in this view of a previously unknown group of galaxies estimated be some three billion light-years away.

LEFT: This image, taken with Hubble's Advanced Camera for Surveys, shows less than a tenth of a starfield in the Constellation of Fornax in the southern hemisphere and was assembled from observations taken between July 2002 and February 2003. The full field which consisted of over 25,000 galaxies was part of an ambitious survey of some of the early parts of the universe known as the Great Observatories Origins Deep Survey (GOODS). The object of the survey was to locate very faint infant galaxies in order to build up a picture of galaxy evolution. The survey located more than 2,000 such examples, providing a considerable amount of new data for the scientists.

FAR LEFT: The Arches star cluster lies only 25,000 light-years away and is the densest known gathering of young stars in our own galaxy, the Milky Way. Although some of the individual stars have been observed and recorded, we cannot get a good overall view of the cluster from earth so data gathered from Hubble has been used to generate the artist's impression of how the cluster would look if viewed from the hub of the Milky Way. Some of the bright blue stars are among the largest found, some 130 times the size of our own sun and mist of the stars are only 2 to 2.5 million years old.

143

LEFT: Despite the great power of the Hubble cameras, it still needs assistance to view some of the most distant part of the universe and such aid is forthcoming from what might be termed a natural zoom lens. Some of the larger galaxy clusters actually distort the path of light traveling through them due to their enormous gravitational fields. Astronomers have learned to harness this effect and in this case some distant objects are viewed through the massive Abell 1689. In fact this technique is a vindication of some of Einstein's work as he predicted that gravity could distort light but did not believe that the effect could ever be observed from earth. It certainly couldn't in his time, but the massive advance that is the Hubble telescope has now provided a clear demonstration of Einstein's great genius.

LEFT: Despite the great power of the Hubble cameras, it still needs assistance to view some of the most distant part of the universe and such aid is forthcoming from what might be termed a natural zoom lens. Some of the larger galaxy clusters actually distort the path of light traveling through them due to their enormous gravitational fields. Astronomers have learned to harness this effect and in this case some distant objects are viewed through the massive Abell 1689. In fact this technique is a vindication of some of Einstein's work as he predicted that gravity could distort light but did not believe that the effect could ever be observed from earth. It certainly couldn't in his time, but the massive advance that is the Hubble telescope has now provided a clear demonstration of Einstein's great genius.

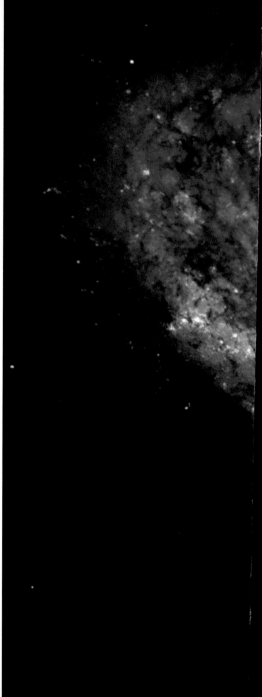

LEFT: The Key Project team used this Hubble telescope view of the NGC4414 spiral galaxy to help calculate the expansion rate of the universe. Based on their discovery and careful brightness measurements of variable stars, astronomers were able to make an accurate determination of the distance to the galaxy. The resulting distance to NGC4414—about 60 million light-years—along with similarly determined distances to other nearby galaxies, contributed to astronomers' overall knowledge of the expansion rate of the cosmos, and helps them determine the age of the universe.

FAR LEFT: This is a view of the G1 globular cluster—the large, bright ball of light in the center of the photograph. G1, also known as Mayall II, orbits the Andromeda Galaxy. Located 130,000 light-years from Andromeda's center, G1 is the brightest globular cluster in the Local Group of galaxies, containing at least 300,000 old stars.

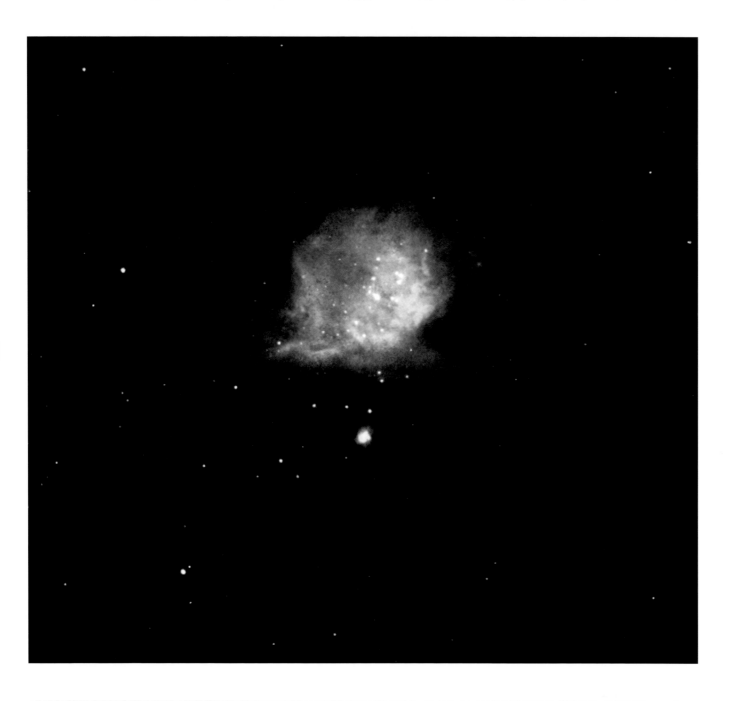

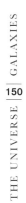

LEFT: This glowing gas cloud is in one of the most active star-forming regions in Galaxy NGC6822. The galaxy lies 1.6 million light-years from Earth in the constellation Sagittarius, one of the Milky Way's closest neighbors. This hotbed of star birth is similar to the fertile regions in the Orion Nebula in our Milky Way Galaxy, but on a vastly greater scale.

RIGHT: This visible-light picture reveals an intergalactic "pipeline" of material flowing between two battered galaxies that bumped into each other about 100 million years ago. The pipeline (the dark string of matter) begins in NGC1410 (the galaxy at left), crosses over 20,000 light-years of intergalactic space, and wraps around NGC1409 (the companion galaxy at right) like a ribbon around a package. The galaxies are some 300 million light-years from Earth in the constellation Taurus.

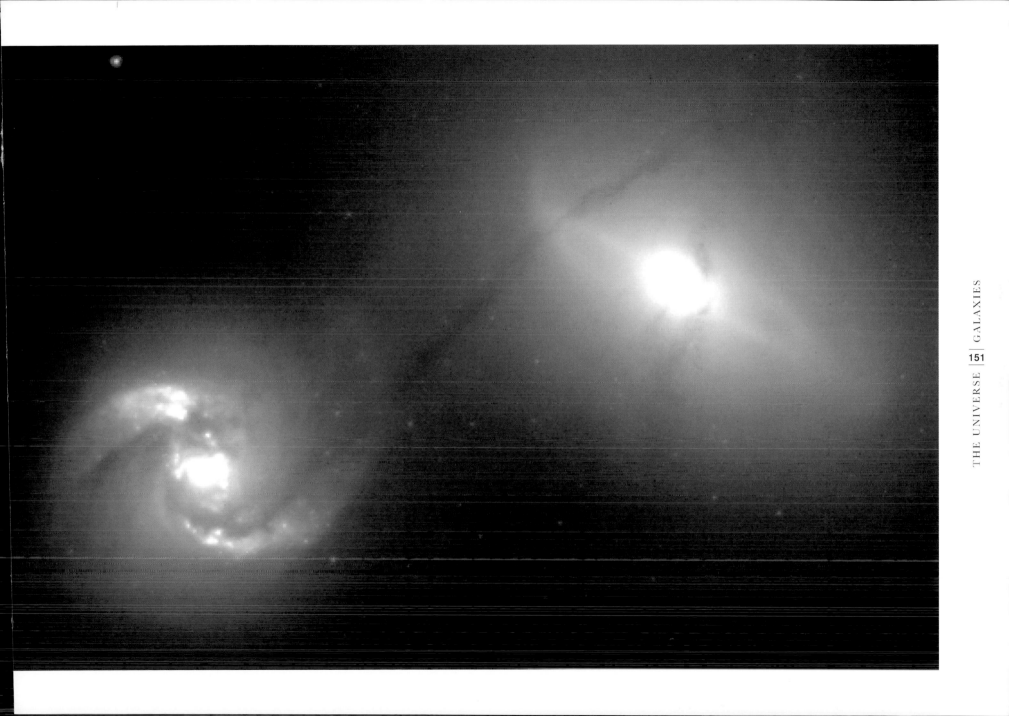

LEFT: In this view of the center of NGC1512, the barred spiral galaxy, the HST reveals a 2,400 light-year-wide circle of infant star clusters. Astronomers generally believe that the giant bar, which is too faint to be seen in this image, funnels the gas to the inner ring, where massive stars are formed within numerous star clusters. Located 30 million light years away, NGC1512 is a neighbor of our Milky Way Galaxy.

FAR LEFT: Most galaxies form new stars at a fairly slow rate, but members of a rare class known as "starburst" galaxies blaze with extremely active star formation. The galaxy NGC3310 is one such starburst galaxy that is forming clusters of new stars at a prodigious rate.

RIGHT: Unusual edge-on view of ESO 510-G13 galaxy, revealing remarkable details of its warped dusty disk and showing how colliding galaxies spawn the formation of new generations of stars. The dust and spiral arms of normal spiral galaxies, like our own Milky Way, appear flat when viewed edge-on. This image shows a galaxy that, by contrast, has an unusual twisted disk structure.

PAGE 156: Our Sun and solar system are embedded in a broad pancake of stars deep within the disk of the Milky Way Galaxy. Even from a distance, it is impossible to see our galaxy's large-scale features other than the disk. The next best thing is to look farther out into the universe at galaxies that are similar in shape and structure to our home galaxy—such as here, spiral galaxy NGC3949.

PAGE 157: A unique peanut-shaped cocoon of dust, called a reflection nebula, surrounds a cluster of young, hot stars in this view. The "double bubble," called N30B, is inside a larger nebula, named DEM L 106. The larger nebula is embedded in the Large Magellanic Cloud, the satellite galaxy of our Milky Way. The wispy filaments of DEM L 106 fill much of the image.

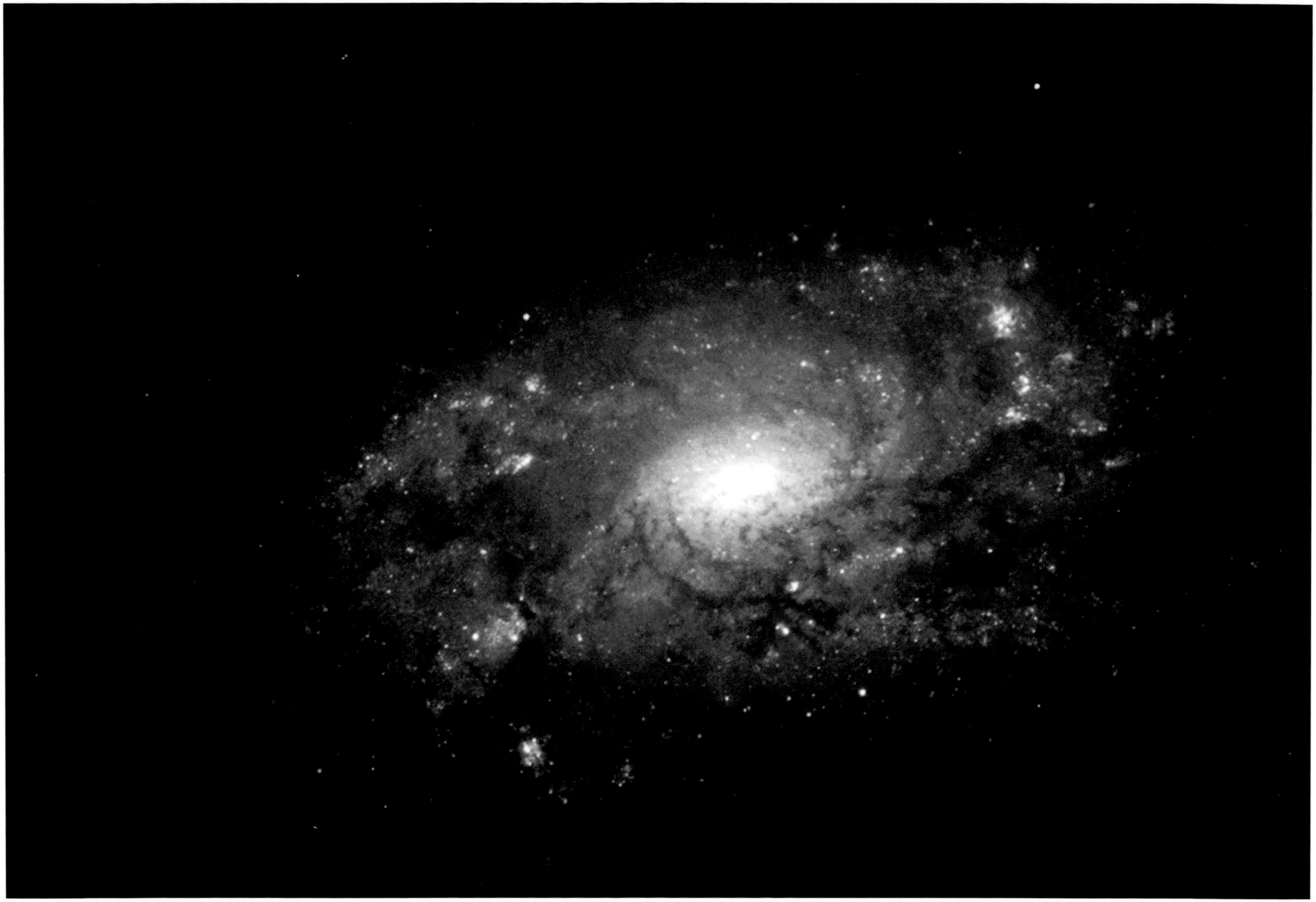

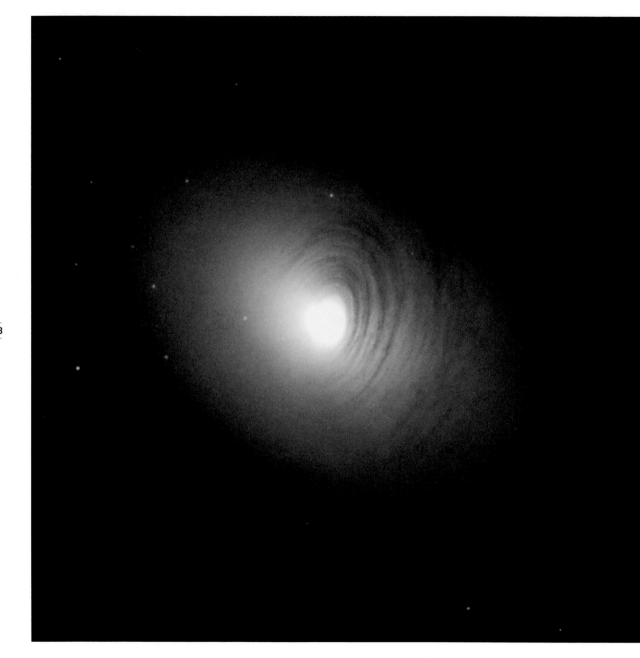

LEFT: Tightly wound, almost concentric arms of dark dust encircle the bright nucleus of Galaxy NGC2787 in the constellation of Ursa Major. Classified as an SB0—a barred lenticular galaxy—this sort of galaxy does not have spiral arms.

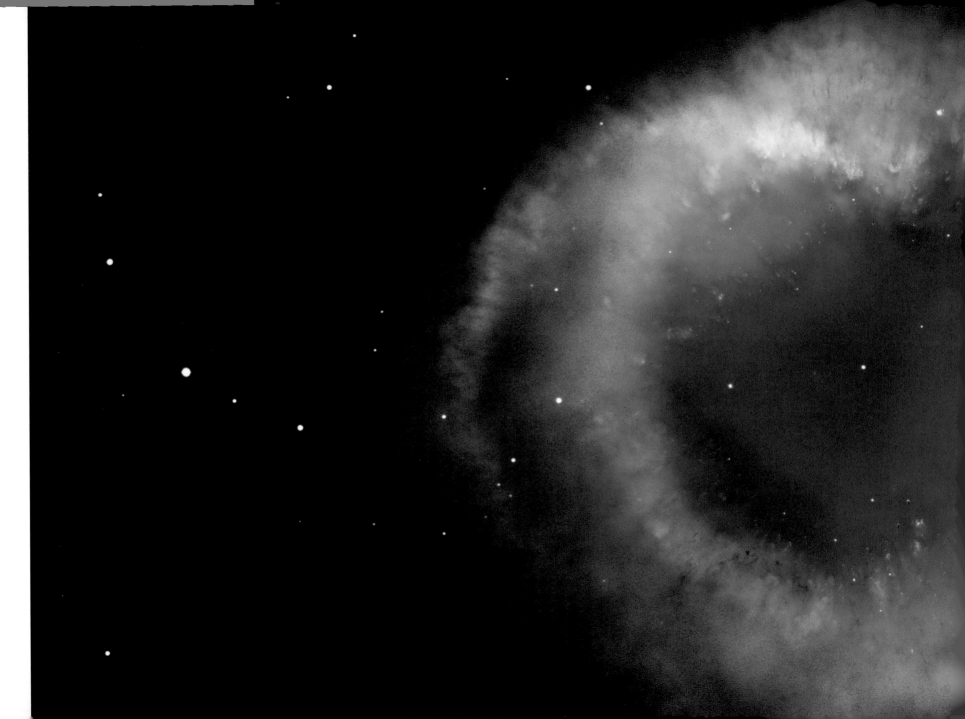

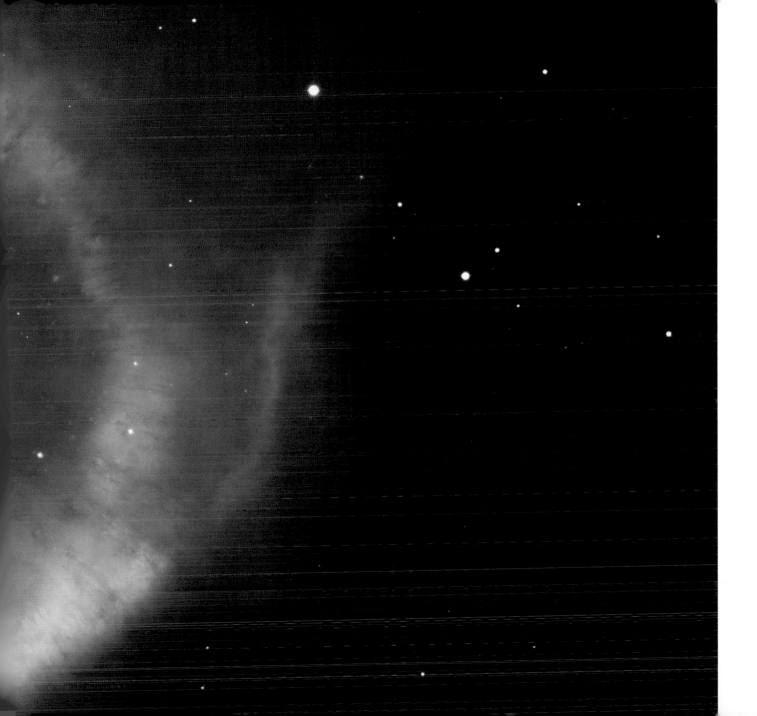

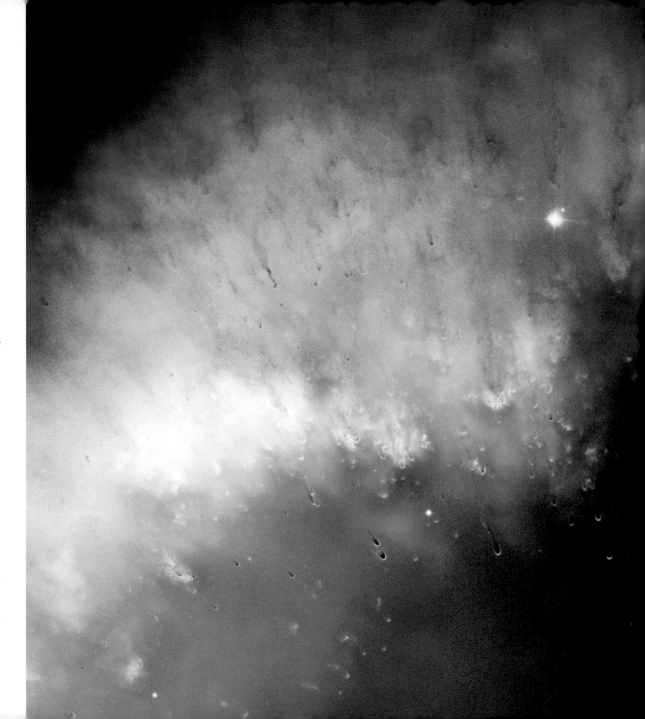

Nebulas

To early astronomers, working without the aid of telescopes, the heavens appear to be filled with stars of various brightness but there were also other objects which appeared larger but less well defined. These were generically termed nebula but subsequent investigation showed that most of these were great concentrations of stars which we now term as galaxies. To the modern astronomer nebula are great clouds of gas and dust which may well contain some stars. Any meaningful investigation of their nature and composition awaited the development of new techniques and instruments such as the spectroscope and infra red photography. This has shown that despite being composed almost entirely of dust and gas, nebula are far from being dormant and in fact are often areas of intense activity and interactions as stars die and others are born, the process forming constant evolution in which the universe constantly recreates itself.

The gas clouds are usually observed as dark areas which block the light from stars farther away. A typical cloud will consist around 90% hydrogen and 9% helium while the remaining 1% is made up of carbon, silicon, oxygen and other trace elements. The single atom element hydrogen is the basic building block of all material in the universe. The dark areas, backlit by stars form dramatic shapes to which mans imagination has applied names so that, for example, there are the Horsehead, Ring and Eskimo nebula. In some nebula the gas clouds have coalesced under extreme gravitational forces to form stars which then burn brightly, heating the surrounding gas which is then driven away. In other cases the surrounding gas begins to radiate its own light as it is heated and this leads to some colourful displays when viewed through the high powered telescopes.

Another form of nebula is the so-called planetary nebula. In fact these are nothing to do with planets and are so named because of their generally circular shape and patterns of colour which makes them resemble a planet when viewed from afar. In fact they are stars which are reaching the end of the lives as they burn off the hydrogen and helium gases. In the process the transform into red giant stars and the outer layers begin to cool although the centre remains hot. As this process accelerates, the star begins to eject dust and gases which form a series of expanding shells around the star, giving rise to the planetary nebula effect.

Other nebula are clouds of dust and gas resulting from cataclysmic events such as a supernova explosion so that they often contain phenomena such as neutron stars, pulsars or even black holes.

The advent of the Hubble telescope has enabled detailed investigations to be made of many nebula and the photographic and digital imaging techniques now available have produced some of the most exciting and awe inspiring views of the universe ever recorded.

LEFT: A detailed closeup of the edge of the Helix Nebula showing comet-like filaments embedded along a portion of the inner rim of the nebula's red and blue gas ring. The nebula is located around only 650 light-years from Earth, making it one of the closest planetary nebulas. Although Hubble imagery was used in creating the complex mosaic picture of which this is part, it was also combined with other images from terrestrial telescopes including the National Science Foundation's 0.9-meter telescope at Kitt Peak National Observatory near Tuscon, Arizona.

THE UNIVERSE | NEBULAS

163

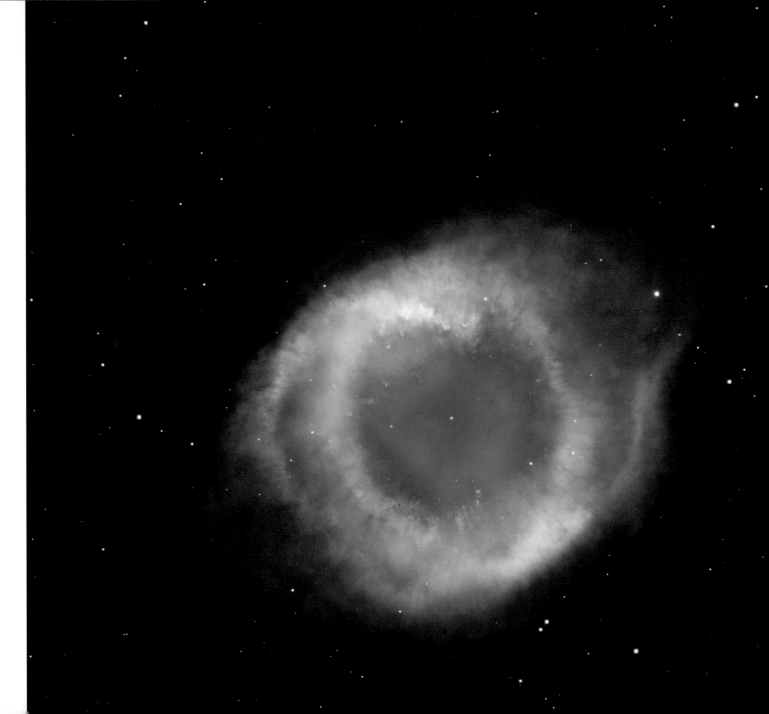

LEFT: The Helix Nebula is a popular target for amateur astronomers and can be easily seen through binoculars as a faint greenish cloud in the constellation of Aquarius, while larger telescopes resolve the ring-shaped nebula. Despite its apparent shape when viewed end-on from Earth, the nebula is actually a cylindrical formation some three billion miles long. This image is the result of several exposures taken on an opportunity basis in November 2002 when the telescope had to be turned away from a meteor storm for half a day. Fortuitously it was then pointing in the direction of the Helix Nebula and a total of nine shots were taken in the process of nine orbits.

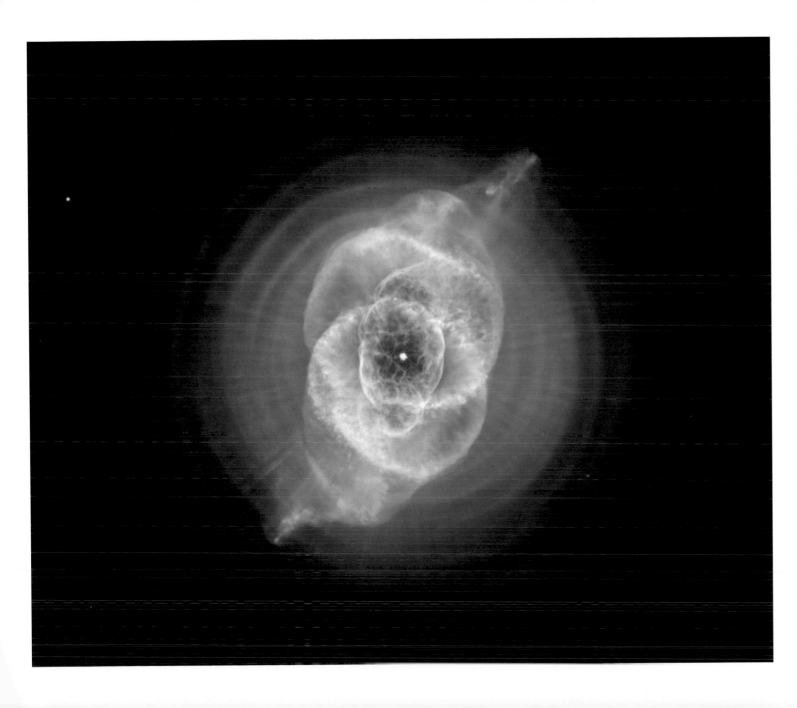

LEFT: The Cat's Eye Nebula (NGC6543) was one of the first planetary nebulas to be discovered; it is one of the most complex such nebulas seen in space. A planetary nebula forms when Sun-like stars gently eject their outer gaseous layers that form bright nebulae with amazing and confounding shapes. In 1994, Hubble first revealed NGC6543's surprisingly intricate structures, including concentric gas shells, jets of high-speed gas, and unusual shock-induced knots of gas. As if the Cat's Eye itself isn't spectacular enough, this new image taken with Hubble's Advanced Camera for Surveys (ACS) reveals the full beauty of a bull's eye pattern of 11 or even more concentric rings, or shells, around the Cat's Eye. Each "ring" is actually the edge of a spherical bubble seen projected onto the sky—that's why it appears bright along its outer edge

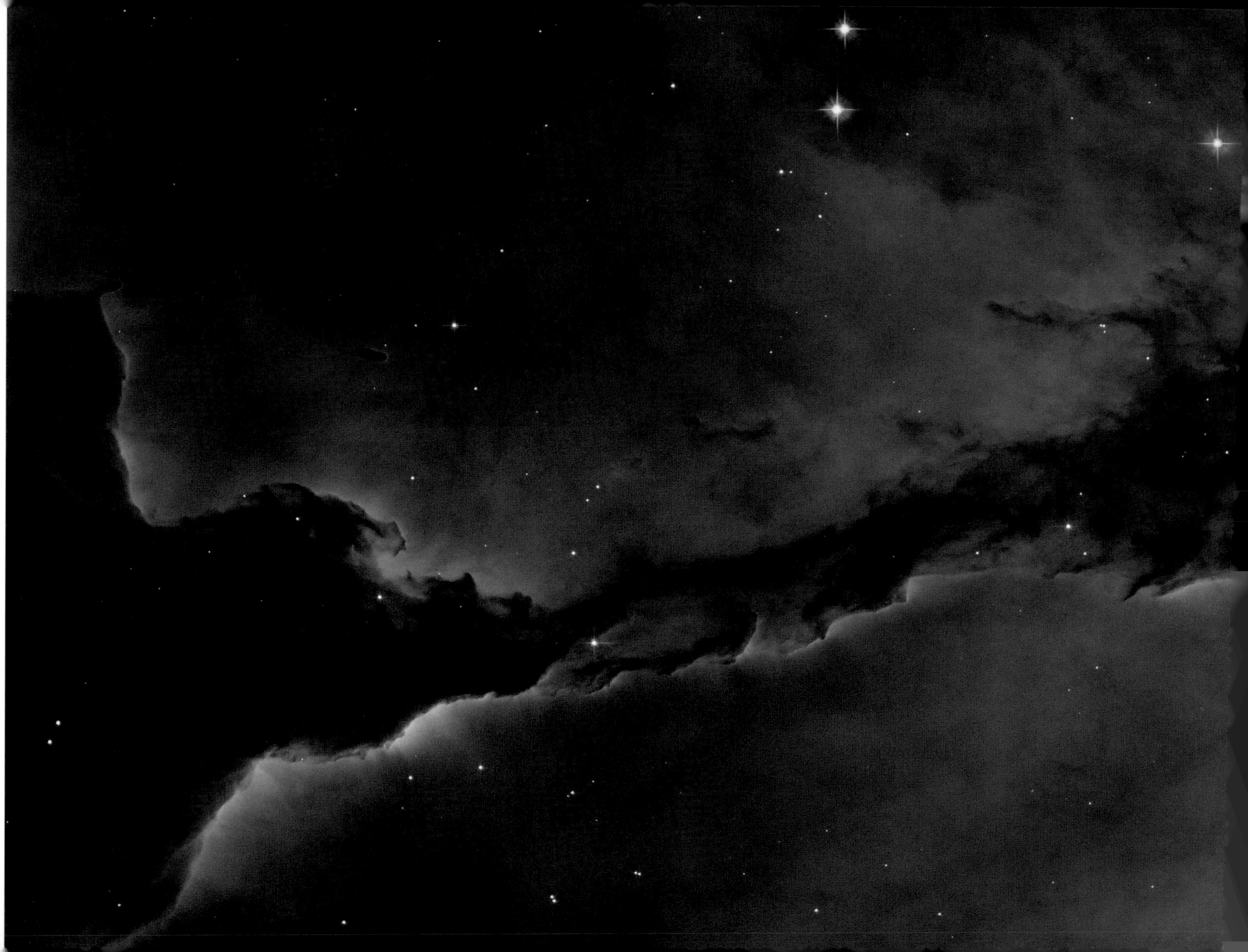

LEFT: This towering column of cold gas and dust rising in the Eagle Nebula is one of the most stunning images ever produced by the Hubble telescope. The total height of this column is 9.5 light years or about twice the distance from the sun to our nearest star. Sometimes this and similar towers in the nebula are called "pillars of creation." The backlight to the plume of gas emanates from a nearby region of massive hot young stars which heats the gases by intense ultraviolet light. This is the first step in a sequence of events as the expanding hot gases compress the colder, dark areas. This image was taken in November 2004 using the Advanced Camera for Surveys.

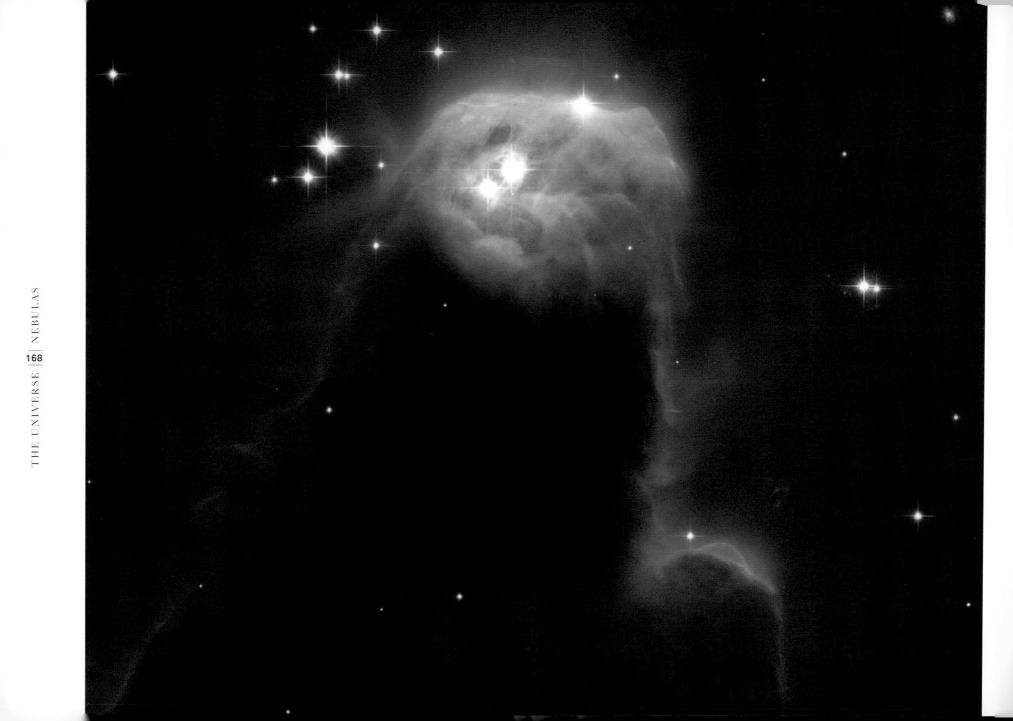

LEFT: This streak of light is known for obvious reasons as the Pencil Nebula (NGC2736) and is located in the southern constellation Vela. Discovered by Sir John Herschel in the 1840s, it is believed to be a result of the huge Vela supernova with the consequential shockwave encountering a region of dense gas, causing it to glow. In this image red light indicates areas of nitrogen gas, green is hydrogen, and blue is oxygen. This was one of the first images obtained following the service mission in 1993 to correct the initial faults in the telescope's primary mirror.

FAR LEFT: Another striking Hubble image, this time showing the Cone Nebula, a similar formation to the Eagle Nebula with tower columns of cold gas and dust forming part of the Swan Nebula (M17), located 5,500 light years away in the constellation of Sagittarius. Photographed in 1995, again using the Advanced Camera for Surveys, the Hubble telescope revealed previously unseen details and has picked out a rich multicolored tapestry of gas clouds which are, in effect, a maternity ward for new stars some of which exhibit signs of embryonic planetary systems.

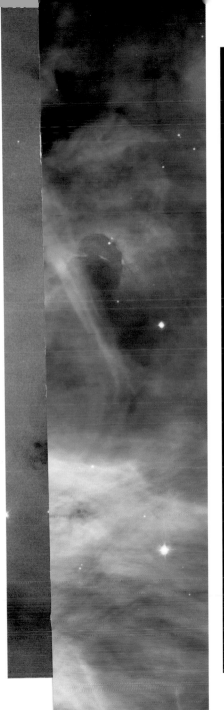

LEFT AND (DETAIL) FAR LEFT: The Swan Nebula, also known as the Omega Nebula. As with other nebulas, its clouds of dark cold gas are roasted under the powerful radiation from adjacent massive stars and the colored areas indicate the presence of various gases including hydrogen, nitrogen, oxygen, and sulphur. On the right of the picture is a rose-like feature, glowing red to indicate hydrogen and sulphur. The darker areas coincide with the more dense pockets of gas which may well contain developing stars. This view was made in April 2002 using four filters (blue, near infrared, hydrogen alpha, doubly ionized oxygen) to construct the color image.

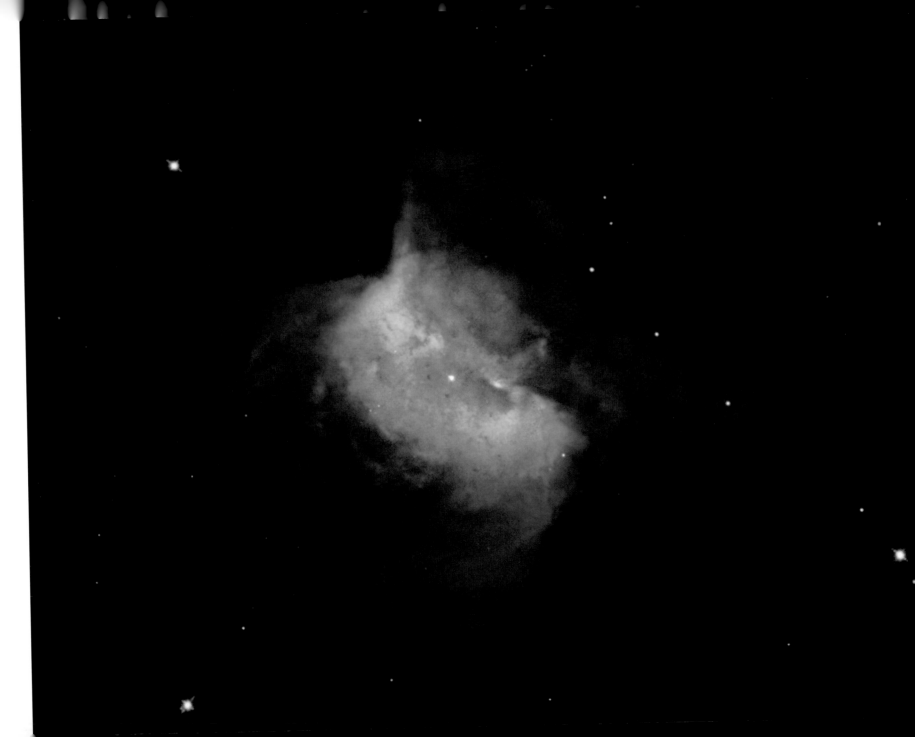

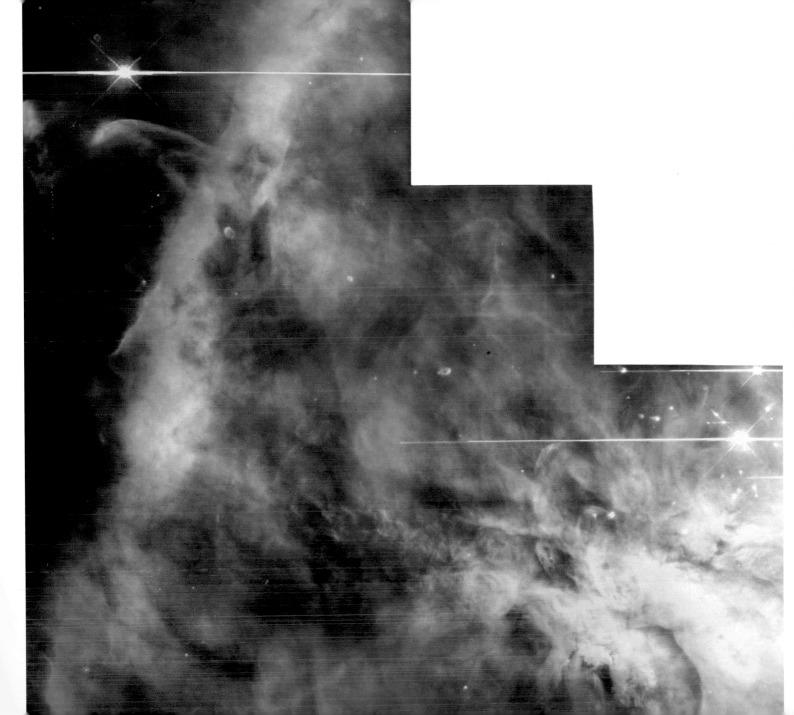

LEFT: A region of the Great Nebula in the constellation of Orion taken by Hubble's Wide Field Planetary Camera 2 in 1994. This is one of the nearest regions of very recent star formation (only 300,000 years ago) and consequently is of great interest to astronomers. The giant gas cloud in the center led scientists to believe that supernova explosion occurred 11,000 years ago and would have been 250 times brighter than the planet Venus, making it a powerful sight in the night sky and even visible by day. Unfortunately for us, this is before the start of recorded history so we have no records of its appearance although it must have been a frightening event to the primitive peoples of the time.

FAR LEFT: This is NGC2346, a so-called planetary nebula which resulted from sun-like stars nearing the end of their lives. It is considered particularly remarkable as its central star is actually a pair of stars in very close proximity orbiting each other every 16 days, a phenomenon known as a binary star. One of these is a red giant star which has pushed out a ring of gas which has then been blown by a stellar wind into the two bubbles or wings which form this dramatic pattern. NGC2346 lies 2,000 light-years away from Earth.

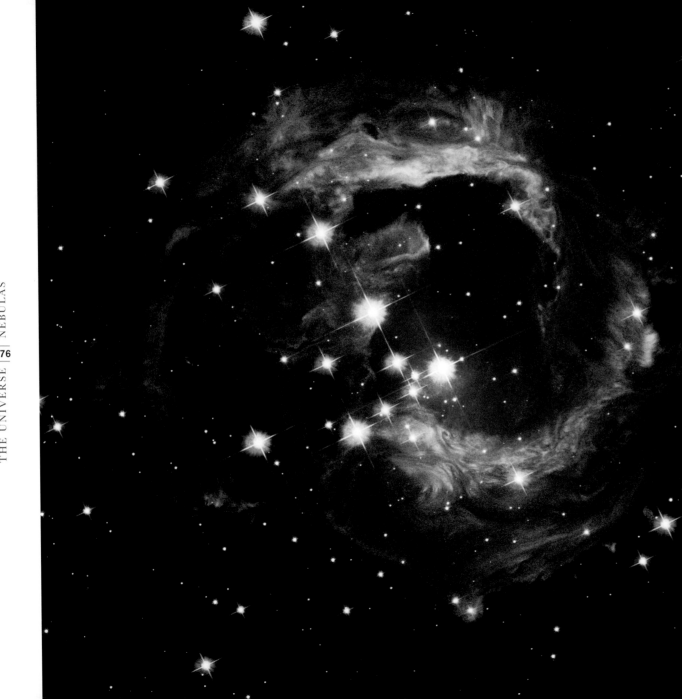

LEFT: Many events in the universe occur over long periods of time, billions of years in some cases, while we can only surmise at the sequence of other events from the evidence visible today. Just occasionally we are able to observe some fantastic event actually as it happens. A case in point is the red supergiant star V838 Monocerotis which lies 20,000 light-years away in the equatorial constellation of Monoceros (The Unicorn). Early in 2002 the star suddenly gave off a massive pulse of light which was visible for several weeks. The star reached a peak luminosity equal to 600,000 times that of our own sun and the flash of light illuminated the dust clouds surrounding it to give this unprecedented image, making the star appear uncannily like a human eye.

Shown here is N49, the remains of a supernova blast set in the Large Magellanic Cloud in the southern skies. The wispy patters of dust and debris hid the fact that N49 contains a pulsar, or neutron star, which is often the result of a supernova explosion. This one rotates at the fantastic rate of once every eight seconds and also has a staggeringly powerful magnetic field, a thousand trillion times greater than that of earth. It is also a prolific generator of Gamma rays, a particularly strong outburst occuring in 1979, and it also emits X-rays. Needless to say, this object is the center of intense study by scientists.

LEFT: A detailed view of part of the Trifid Nebula showing a small part of a dense cloud of dust and gas, some eight light-years away from the center of the nebula. A thin jet of dust protrudes from the head of the cloud, rather like an antenna on an insect's head. It is actually formed by gases escaping from the formation of new star inside the main cloud and it is almost one light-year long. In absolute terms, the life of this star is not likely to be very long. Over the next 10,000 years the glare of heat and radiation from the massive central star will erode the nebula and overrun the nascent star, bringing its development to a premature end.

RIGHT: A strangely ephemeral image showing an opaque dust cloud silhouetted against the bright stars of an area where new stars are constantly forming known as IC2944. This is relatively close to earth at 5,900 light-years away in the constellation of Centaurus. Such dense clouds of gas are known as globules and were first noticed by the Dutch-American astronomer Bart Bok in 1947. They appear to view because the less dense areas around them have been burned away by the radiation of the cluster of O-type stars (the hottest) in the region. Ultimately they will suffer a similar fate but had there not been this radiation then they may themselves have collapsed to form low-mass stars such as our own sun.

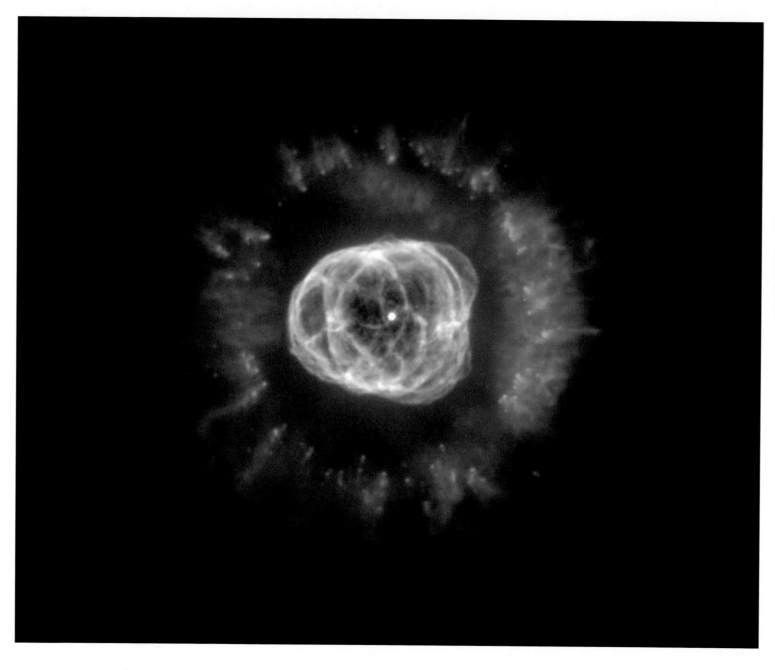

LEFT: This is the Eskimo Nebula (NGC2392), so called because of its likeness to a face wrapped in a fur parka hood. In fact it is a planetary nebula, the remains of a dying sun-like star which began forming some 10,000 years ago as the dying star threw out debris and dust. It was first recorded as long ago as 1987 by William Herschel and is quite close to us, being some 5,000 light-years away in the constellation of Gemini. This image was taken in January 2000 and the gases nitrogen, hydrogen, oxygen, and helium are represented by the colors red, green, blue, and violet respectively.

RIGHT: Another feature to be found in the Large Magellanic Cloud is the 30 Doradus Nebula. Its most spectacular feature is the star cluster R136 (the bright area left of center) which is responsible for the overpowering ultraviolet radiation reacting with the enveloping clouds of dust and gas to form a fertile region for further star formation. One classic feature is the existence of pillars of dense dust, first associated with the Eagle Nebula and astronomers now realize that these are areas within which stars are formed. The inner region of 30 Doradus measures approximately 200 by 150 light-years and the whole is 170,000 light-years from Earth.

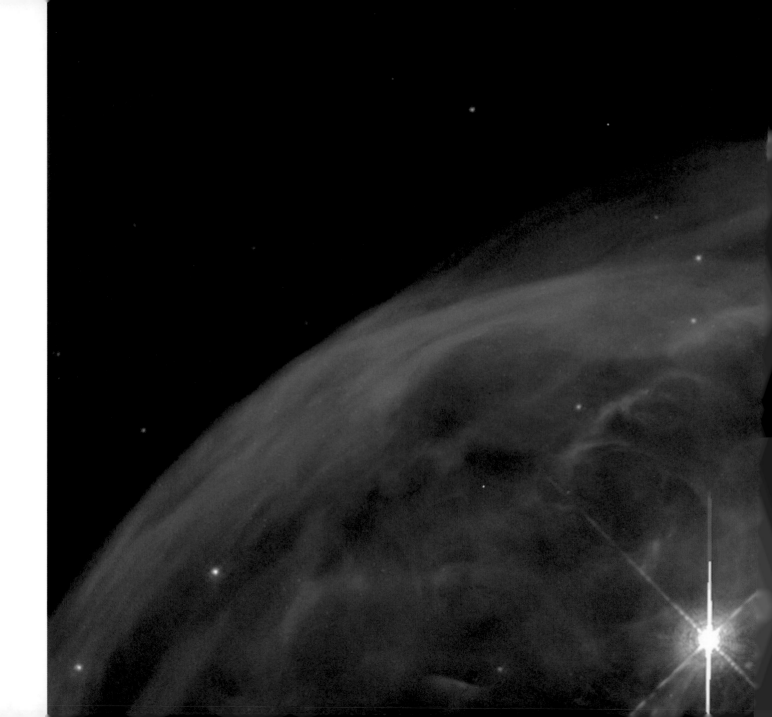

RIGHT: The bright star at lower center in this Hubble image taken in 2000 is 40 times more massive than the sun. Its heat is causing a giant bubble of gas and dust to expand around it forming the eponymous Bubble Nebula which is currently six light-years wide and is expanding at the rate of four million miles per hour. It is situated in the constellation Cassiopeia, some 7,100 light-years from earth. Cassiopeia is one of the easier constellations to spot, its distinctive W shape being high up in the northern sky.

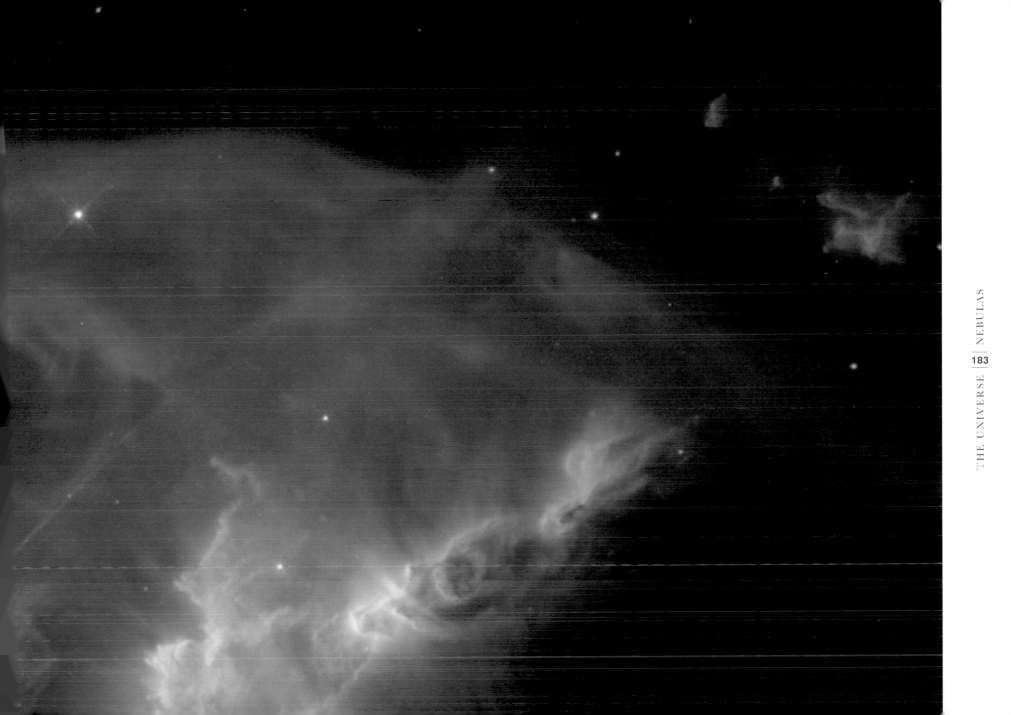

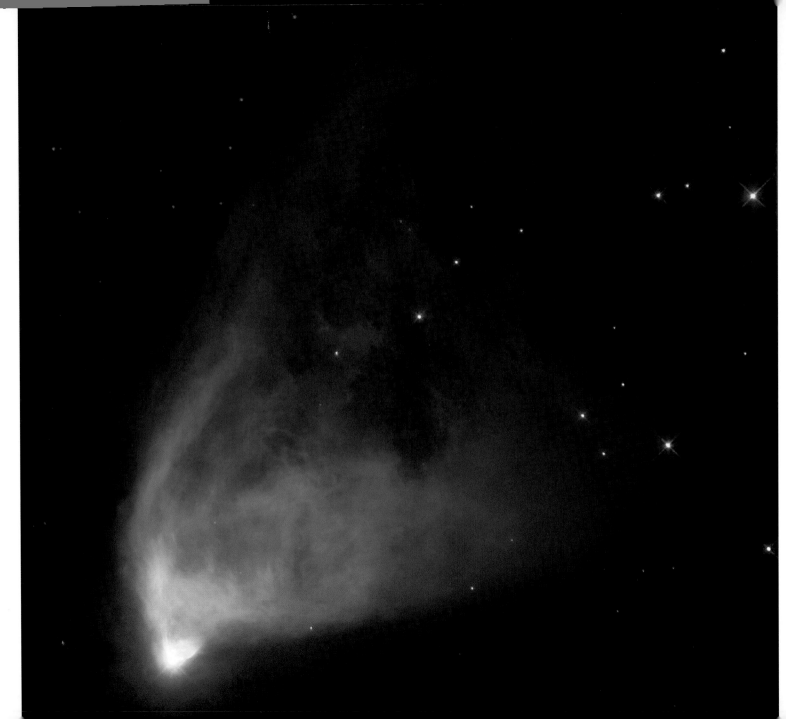

LEFT: The American astronomer Edwin Hubble did not only have a telescope named after him. He was a prominent observer and among many objects which he discovered and catalogued is this formation known as Hubble's Variable Nebula (NGC2261). It is actually a fan-shaped cloud of gas and dust illuminated by the star R Monocerotis, the bright star showing at the lower edge of the nebula. In fact the star itself cannot be directly observed due to the layer of dust between it and ourselves. It lies on the edge of the Milky Way, 2,500 light-years from earth and is only 300,000 years old with a mass around ten times that of the sun.

RIGHT: A planetary nebula, IC4406 formed from the debris of a dying star. In this instance we are viewing it from its side but viewed from another angle it would appear as a doughnut ring with the dying star in its center. As it is, the fine tendrils of material look like the veins in the retina of a human eye and consequently this object is known as the Retina Nebula.

PAGE 186: This strange object is a proto-planetary nebula and represents an early stage in the formation of planetary nebula such as the Ring and Eskimo nebulas. When stars similar to the sun begin to burn out, they expand to become red giants: their outer layers are thrown into space. Subsequently the hot core causes the dust and gases to glow, forming the nebula. However in the initial stage the core is not uncovered and the gases do not glow so brightly. In this case the nebula is viewed edge on, and an expanding illuminated field of dust and gases is obscured by an outer ring of dark dust, appearing to cut it in half. First discovered on sky photos taken by Aturo Gomez, a Chilean astronomer, it has inevitably become known as Gomez's Hamburger. It lies 6,500 light-years away in the constellation of Sagittarius.

PAGE 187: Another object to be found in the constellation Cassiopeia is this riot of colorful streamers of dust and gas, the remains of one of the biggest supernova explosions to be observed in recorded history, being seen from earth in the late seventeenth century. The star which created the supernova was around 20 times the size of our sun but, as such, was fated to have a much shorter life. These large stars use up their supply of nuclear fuels at a much greater rate, leading eventually to a rapid collapse of the core and creating the enormous gravitational forces which rip the star apart. Such is the nature of the universe however, that these cooling knots of gas may well be recycled into new stars in the future. Own sun and solar system was created in exactly this way billions of years ago.

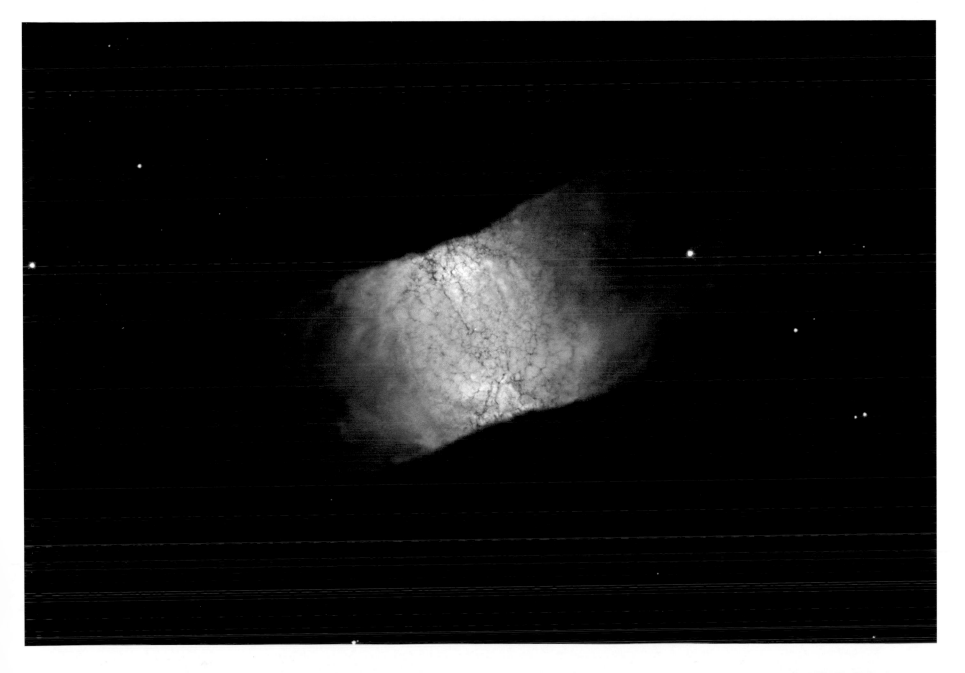

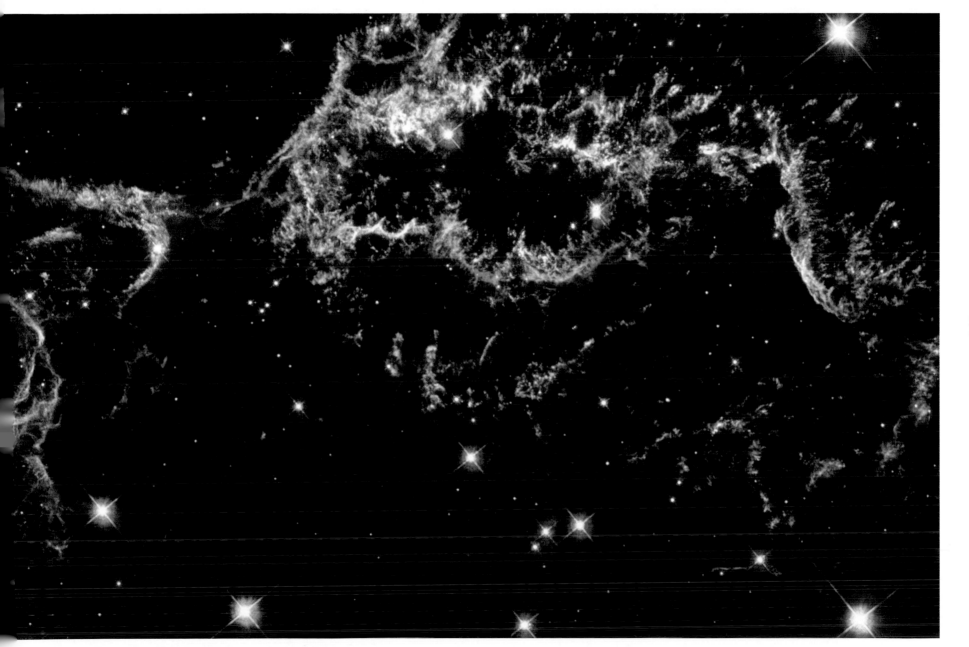

FAR LEFT: An image of N11 in the Large Magellanic Cloud, taken with the Curtis Schmidt telescope at Cerro Tololo Interamerican Observatory (CTIO). This color image was produced using three separate exposures taken in hydrogen, sulfur, and oxygen bandpasses. The green outline indicates the extent of the image made by Hubble.

LEFT: Star birth in a neighboring galaxy. The star-forming region N11B lies in the Large Magellanic Cloud. The blue and white-colored stars near the left edge of the image are among the most massive stars known anywhere in the universe. Another generation of stars is being created inside the dark dust clouds in the center and right-hand side of the image.

LEFT: Rising from a sea of dust and gas like a giant seahorse, the Horsehead Nebula is one of the most photographed objects in the sky. Also known as Barnard 33, the Horsehead is a cold dark area of dust and gas silhouetted against the bright nebula IC434. The unusual shape was first noted in a photographic plate made in the late nineteenth century. This distinctively shaped nebula is a target for adventurous amateur astronomers. Located just south of the bright star Eta Orionis, the left hand of the three forming Orion's belt, it is just possible to distinguish it with a typical commercially available telescope.

RIGHT: This photograph, taken by the Hubble Wide Field Planetary Camera 2, was taken in May 1999 and shows a portion of the Swan Nebula in the constellation Sagittarius. It shows a bubbly ocean of glowing hydrogen gas, together with small amounts of other elements including oxygen and sulphur. The wave-like patterns are caused by constant stream of ultraviolet radiation from nearby young massive stars which causes the heated surface of the dense hydrogen clouds to glow orange and red creating a spectacular effect. Almost inevitably this process will result in the formation of new stars.

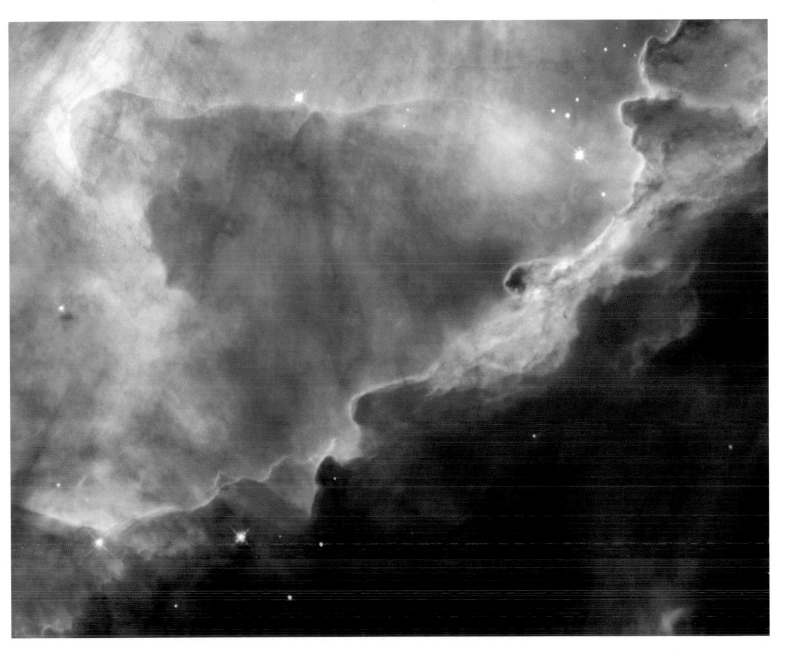

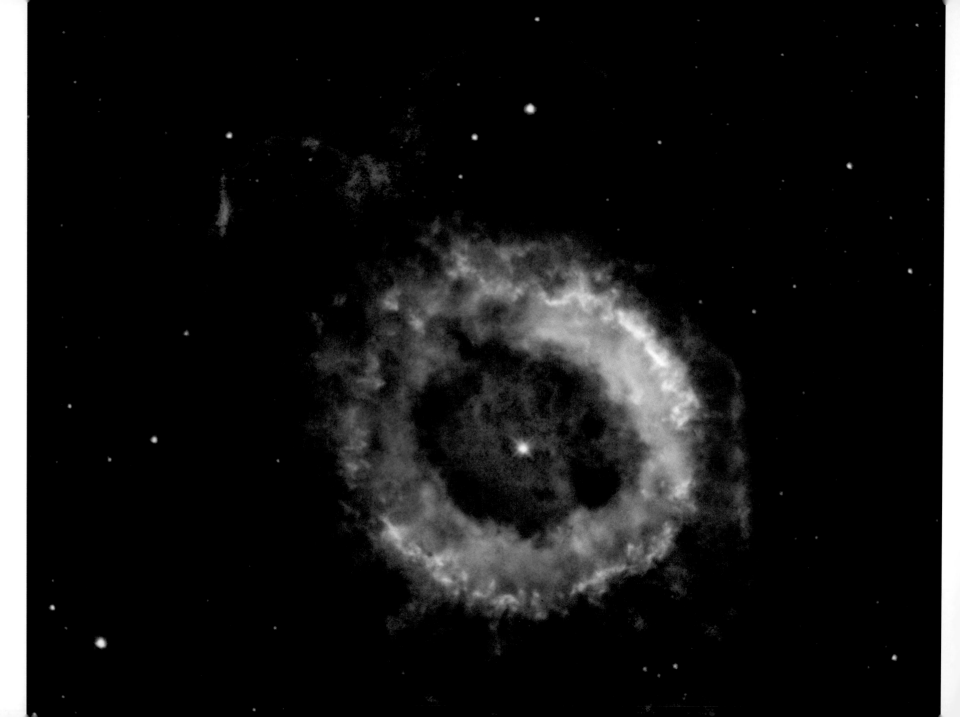

LEFT: One of the early images taken with the Hubble Wide Field Camera following its launch in 1990, this shows an area at the edge of the Great Nebula in Orion, some 1,500 light-years from earth. Despite the spherical aberration which caused problems at the time, this image was still far superior to anything seen by terrestrial telescopes. The scientific data obtained from analyzing the multicolor gas clouds was considerably in advance of previous data and helped the ongoing investigations into the origins and evolution of the universe. The red areas are gaseous sulphur while clouds of oxygen and hydrogen are clearly differentiated in blue and green respectively.

FAR LEFT: This planetary nebula known as the Little Ghost Nebula (NGC6396), which lies in the direction of the southern constellation Ophiuchus (The Serpent Holder), is thought to be between 2,000 and 5,000 light-years away. As with other examples, this is a dying star which is discarding dust and debris as i burns out and breaks up. The prominent blue-green ring marks the region where high energy UV light has stripped electrons from the atoms of the hydrogen gas in a process of ionization. The outer red areas are where nitrogen gas atoms have also been ionized. This picture shows how our own sun may look in the future, some five billion years hence.

RIGHT: The Hubble telescope is capable of resolving phenomena which could never be seen from Earth, both because of the increased clarity of images obtained from space and also because of the sophisticated filters which can be applied to the onboard cameras. A striking example is the hypersonic shockwave formed by material moving at an incredible 148,000 miles per hour in the Orion Nebula, 1,500 light-years away. It is likely that such shockwaves accompany the ejected material from a newly formed star and in this case the plume is only 1,500 years old

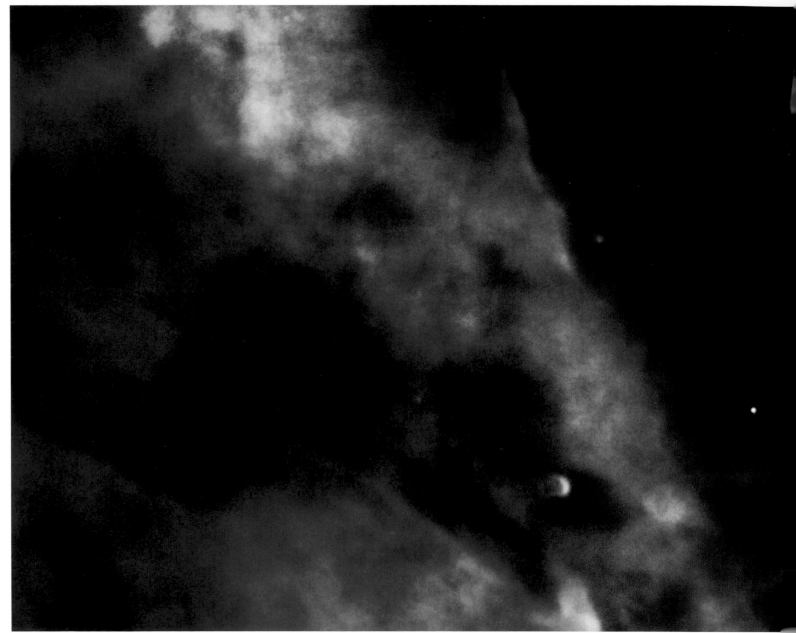

RIGHT: The bright white dot in the center of Nebula 2440 shown here is one of the hottest known stars. Based on observations through the Hubble telescope, astronomers calculate that the star has a temperature of at least 200,000°C (360,000°F). Prior to Hubble, terrestrial telescopes were unable to offer the necessary degree of discrimination to separate the star from the nebula background or to appreciate the intricate structure of the nebula with its filaments and streamers of dust and gas.

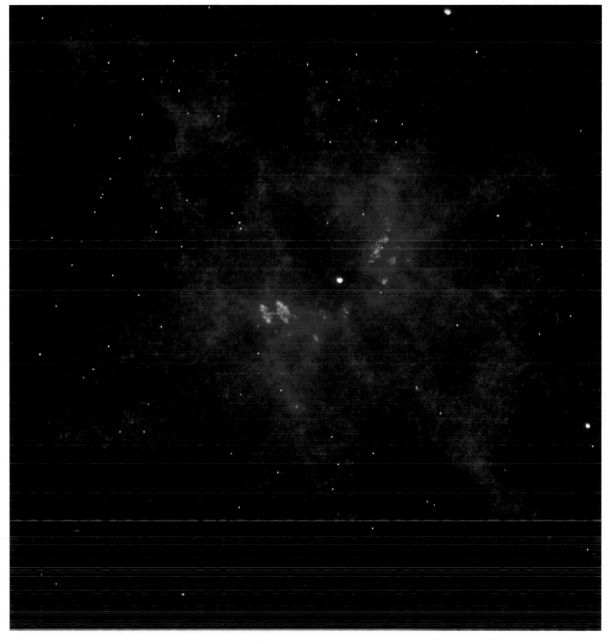

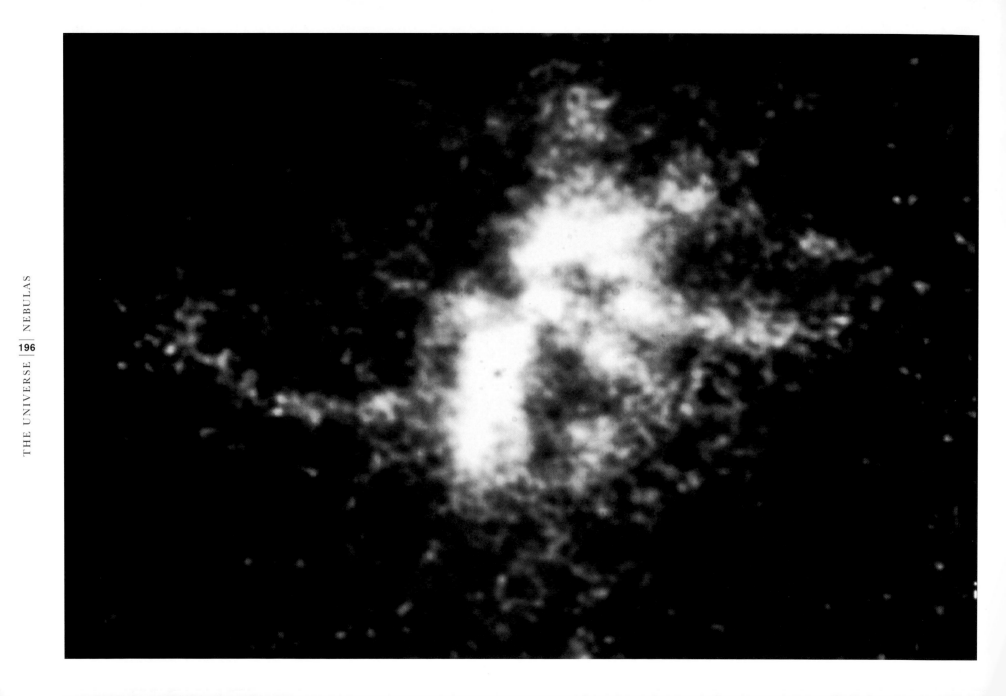

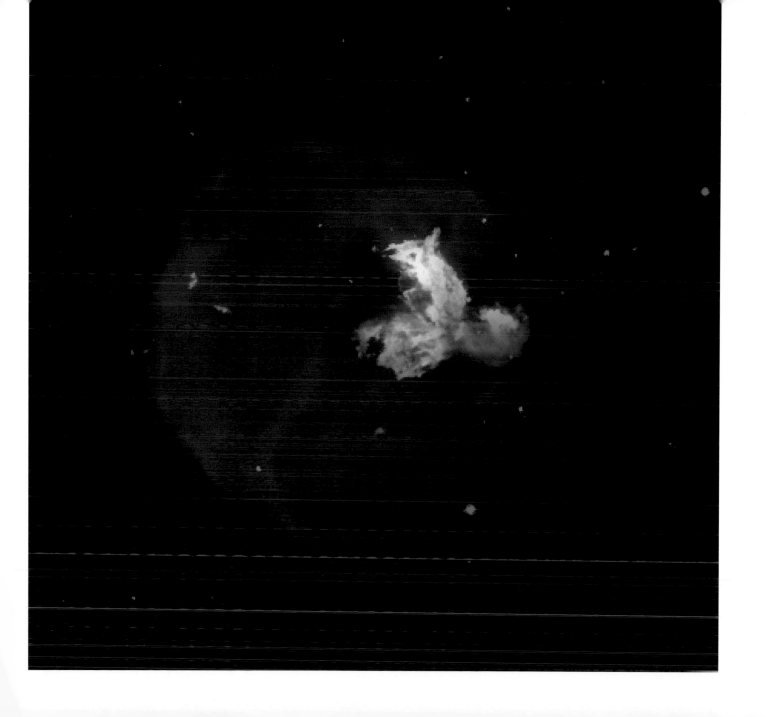

LEFT: A supernova remnant, LMC N63A, viewed through the X-ray filter of one of Hubble's cameras. The X-ray glow, defined in blue, results from material heated to 10 million °C (18 million °F) by a shockwave generated by the supernova explosion. Optical visible traces are shown by the green areas while the red areas are produced by radio emissions from the blast. This composite picture shows the wealth of details which can be obtained from a single object by the use of different sensors.

FAR LEFT: The Large Magellanic Cloud in the southern hemisphere is home to a number of interesting objects including this irregular planetary nebula N66. The image was actually obtained in June 1991 using the European Space Agency's Faint Object Camera. This was the first time that a planetary nebula outside the Milky Way (it lies 169,000 light-years away) had been seen so clearly. The nebula was ejected by a luminous giant red star which subsequently contracted to form a blue remnant star located at the center of this image.

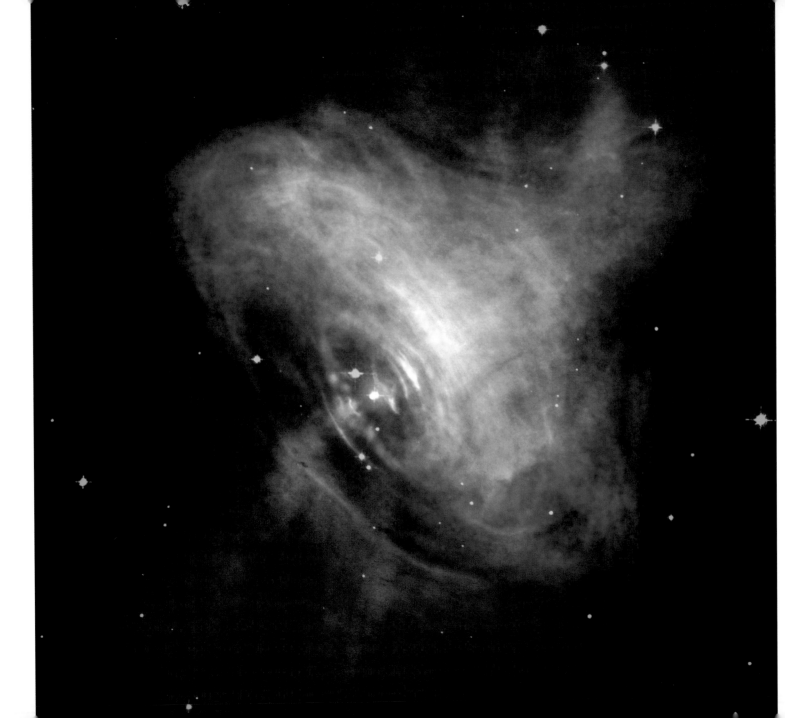

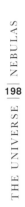

LEFT: Another composite picture showing X-ray (blue) and optical (red) images combined to achieve a dramatic effect. In this case the target is the Crab Nebula situated in the constellation Taurus which was always well known for its high energy radiation, although the cause of this was not explained until pulsars were discovered by British radio astronomers in 1967. Pulsars are high-density neutron stars of very small size and the one known to be at the center of the Crab Nebula was formed following a well-documented supernova explosion observed by Chinese astronomers in 1054 A.D.

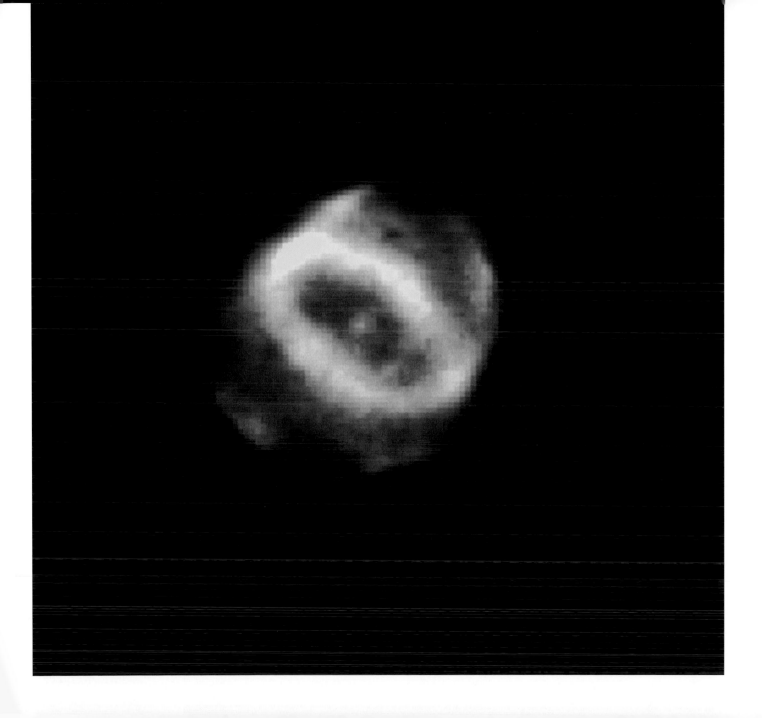

LEFT: A Hubble telescope image of the recently formed planetary nebula Hen 1357, so-called because it was the 1,357th object in a list of unusual stars complied by the astronomer Karl Henize. It is situated 18,000 light-years away in the southern constellation of Ara (the Altar) and is a good example of nebula about which little was known prior to the advent of the Space Telescope as terrestrial telescopes were unable to resolve enough detail. The prominent yellow ring is a concentration of gas ejected by the aging star at the center and appears to be tilted some 35 degrees from the horizontal as viewed from Earth.

LEFT: The Cygnus Loop Nebula appears a faint ring of glowing gases in the northern constellation of Cygnus (the Swan). As with most nebulas, it is a supernova remnant from an enormous stellar explosion, in this case one which occurred about 15,000 years ago, and is some 2,600 light-years away within the plane of the Milky Way. This picture homes in on a small section of the ring where the outwardly spreading shockwaves from the blast have compressed and heated the gases, causing them to glow. The bluish ribbon of light is thought to be a knot of gas ejected by the supernova and travelling at three million miles per hour. Of technical interest is the fact that the blue areas show oxygen heated to between 30,000 and 60,000°C (54–108,000°F), while the red areas are sulphur gas at temperatures around 10,000°C (18,000°F).

LEFT: A detailed view of part of the Helix Nebula, the closest planetary nebula to earth at distance of 450 light-years. It shows an area where two clouds of gas have collided near a dying star, the interaction producing numerous bright tadpole-shaped objects called "cometary knots" because they look like small comets with their tails. Although they appear small in this picture, in fact the "head" of each knot is twice the size of our solar system and the "tail" is over 100 billion miles long. Such phenomena have been observed elsewhere but not on the scale seen in the Helix Nebula where they number several thousand.

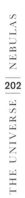

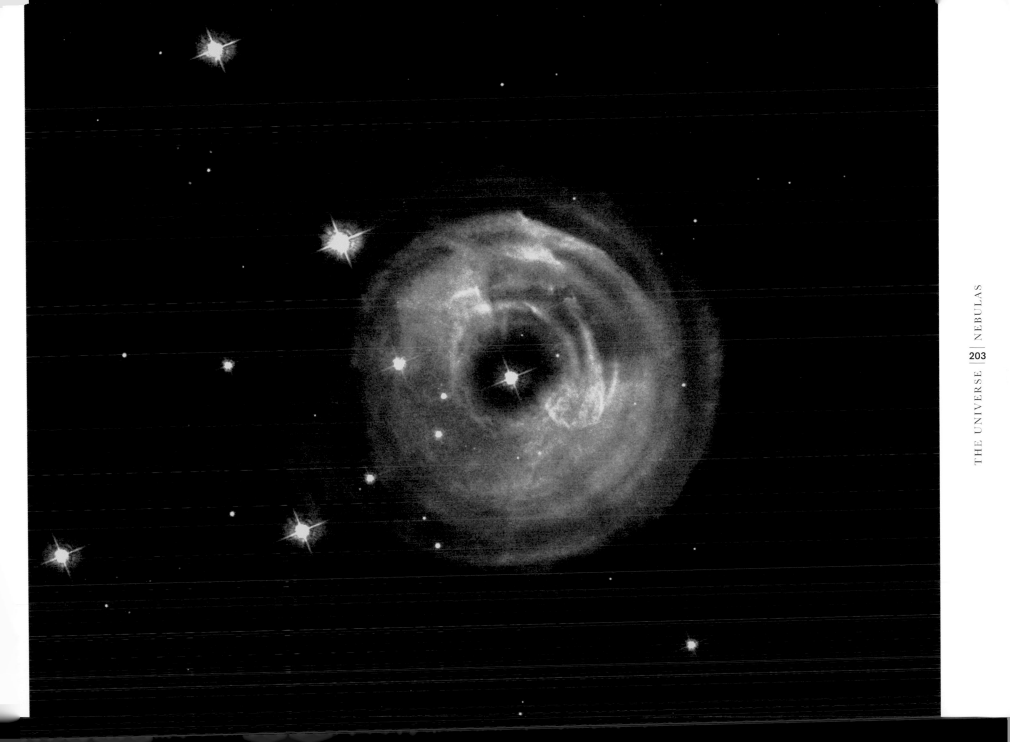

LEFT: Another view of the Cygnus Loop (see page 200) showing further effects of the supernova blast wave which is moving from left to right across the field of view causing a dense cloud of interstellar gas to glow as it is heated by the impact. Although astronomers were aware of the shockwave effect in nebulas, it was not possible to study them in any detail until the advent of the Hubble Space Telescope which has had the same impact on space science as the microscope did for the study of the human body.

FAR LEFT: A detailed montage showing a sweeping view of the 30 Doradus Nebula. The R136 star cluster is at the center of this nebula, providing the heat and energy to heat and illuminate the massive clouds of dust and gas which make up the nebula. Because of Hubble's high resolution, it is often necessary to compose mosaics such as this to show the whole picture of an area of interest. Digital-imaging techniques are essential to achieve the necessary accuracy in the blending of adjacent areas..

FAR LEFT: The Eagle Nebula (see page 166) provides some of the most fantastic images yet seen in deep space. In the early days of modern astronomy it was designated M16, the 16th image in Charles Messier's eighteenth-century catalogue of "fuzzy" objects which were obviously not comets as they remained fixed in space. Viewed by means of the Hubble Wide Field and Planetary Camera 2, its most prominent features are the huge pillars of dust and gas which are denser than surrounding areas and therefore slower to erode under the intense ultraviolet radiation of nearby bright stars. However, as the pillars themselves are eroded, smaller globules of even denser gas will remain, incubators to a new generation of stars as the universe constantly regenerates.

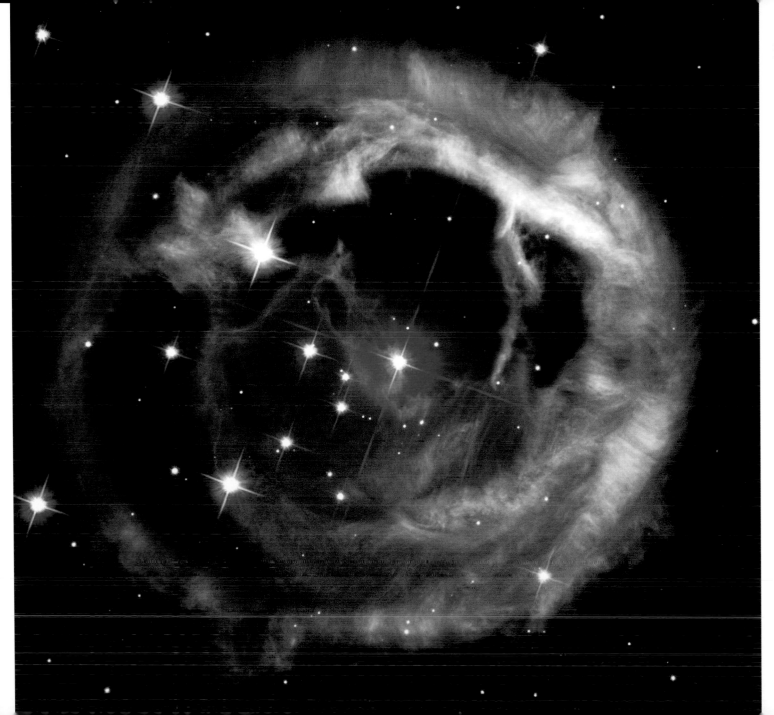

LEFT: Another view of V838 Monocerotis (see pages 176 and 203).

LEFT: A composite view of part of the Crescent Nebula (NGC 6888) in the northern hemisphere constellation Cygnus (the Swan), over 4,700 light-years from earth. At the center of the nebula is a rare and short-lived class of super-hot star (WR136) called a Wolf-Rayet, which is responsible for a fierce stellar wind of charged particles flowing outward, forming shockwaves which react with the gases surrounding the nebula so that they become visible in this striking range of colors. This also confirms the presence of previously unseen matter around the nebula, explaining the difference between the amount of matter lost from WR136 as a red supergiant and the much smaller amount of matter observed until these 1995 images became available.

RIGHT: Another of the objects cataloged by Charles Messier, this is M8—the Lagoon Nebula—which lies 5,000 light-years away in the constellation of Sagittarius. Just visible at top left is the very hot type O star, Herschel 36, which is the main source of ionizing radiation for the brightest part of the nebula, which is known as the Hourglass. The ionizing radiation induces a process known as photoevaporation in the surfaces of the gas and dust clouds, and deflects the violent stellar winds which are tearing into the cooler clouds. The outcome is a series of spiral shapes analogous to tornadoes (twisters) on earth.

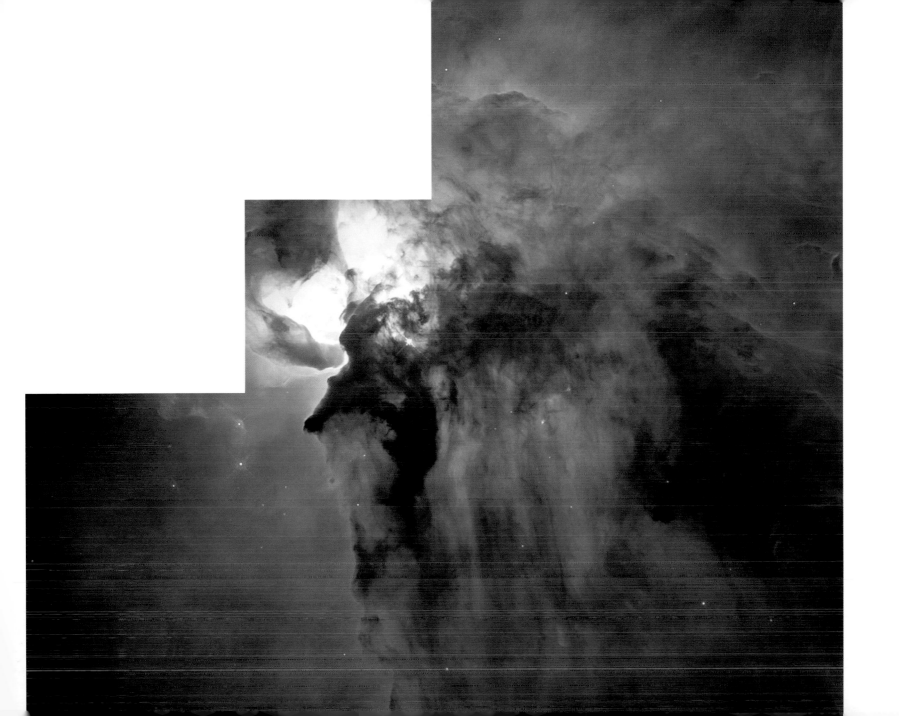

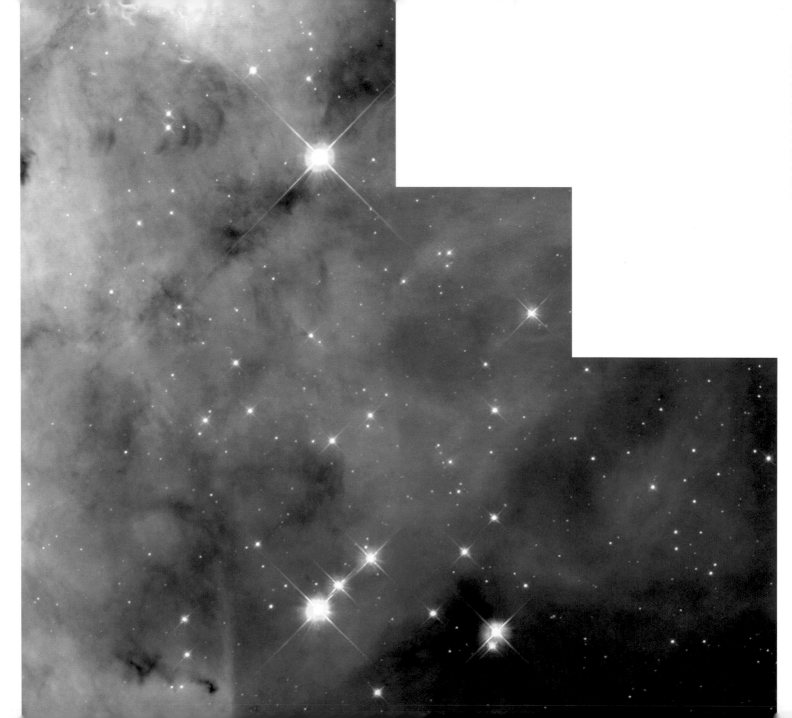

LEFT: As already mentioned, the Large Magellanic Cloud in the southern sky is home to the 30 Doradus Nebula but it also contains another large star-forming nebula designated N11, part of which is shown here. In the upper left of the picture are a collection of blue and white Type O and B stars which are extremely hot. These examples are among the most massive stars known anywhere in the universe. To the right are dense, dark, dust clouds which, due to the radiation from the neighboring stars, are beginning to heat up and energize, and may in time form further new stars. This process is known as sequential star formation and an almost perfect example is shown here.

LEFT: Although catalogued as M27 by
Charles Messier, the Dumbbell Nebula
shown here was in fact the first planetary
nebula ever discovered, in 1764. This
closeup view of a section of the nebula
shows intriguing knots of gas which have
formed at the interface of the hot ionized
and cool neutral gases. Each ranges in size
from 11 to 35 billion miles across, although
their total mass is only equivalent to three
Earths. Similar knots have been found in
almost all other planetary nebula including
the Retina and Eskimo nebulas illustrated
elsewhere.

RIGHT: A general view of the Carina Nebula (see also page 191), which contains the ultraluminous variable star Eta Carinae, visible at the top of this mosaic of the nebula. Once again the patterns formed by gas and dust clouds reacting to the power of the ultraviolet radiation from the bright stars have produced a spectacular display which is, literally, out of this world. This image was produced as part of a project which used several of the Hubble's instruments to simultaneously observe the nebula. These included the Space Telescope Imaging Spectrograph and the Wide Field Planetary Camera 2.

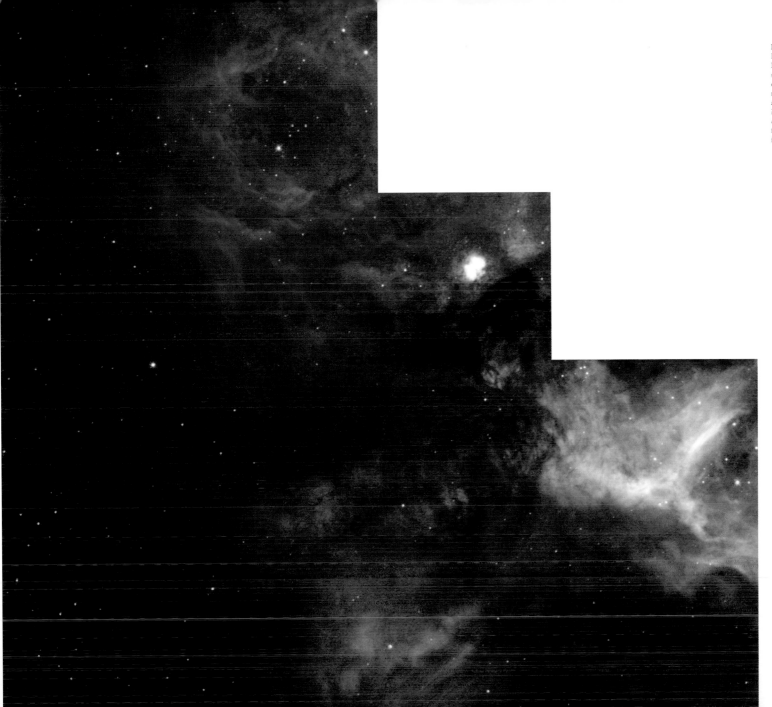

LEFT: Yet another object from the fertile Large Magellanic Cloud. This is the Papillon Nebula (N1590, so called because of its supposed butterfly shape formed by two gas clouds flowing from the formation of a new massive star. It is thought that such new stars are so hot that radiation pressure halts the infall of gas and directs it away in opposite directions. A disk of denser matter trying to fall onto the star acts as a focusing ring to force the gases into a bipolar stream.

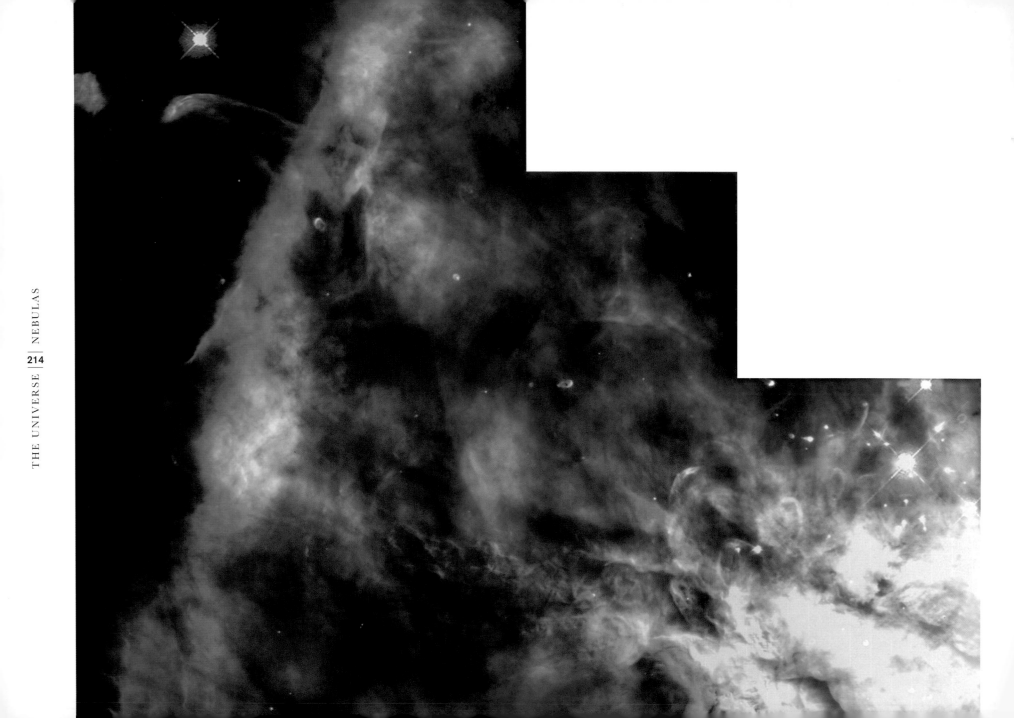

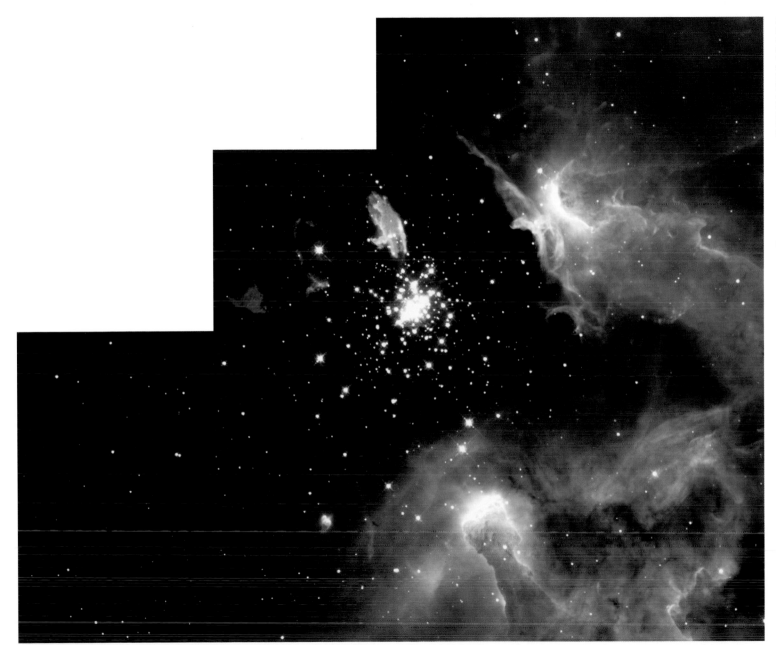

LEFT: In this stunning picture of the giant galactic nebula NGC3603, the Hubble telescope's crisp resolution captures various stages of the life-cycle of stars in one single view. This picture nicely illustrates the entire stellar life-cycle of stars, starting with the Bok globules and giant gaseous pillars (evidence of embryonic stars), followed by circumstellar disks around young stars, and progressing to aging, massive stars in a young starburst cluster. The blue super-giant with its ring and bipolar outflow (upper left of center) marks the end of the life-cycle.

FAR LEFT: Another view of the great Orion Nebula. (See also pages 172 and 175.)

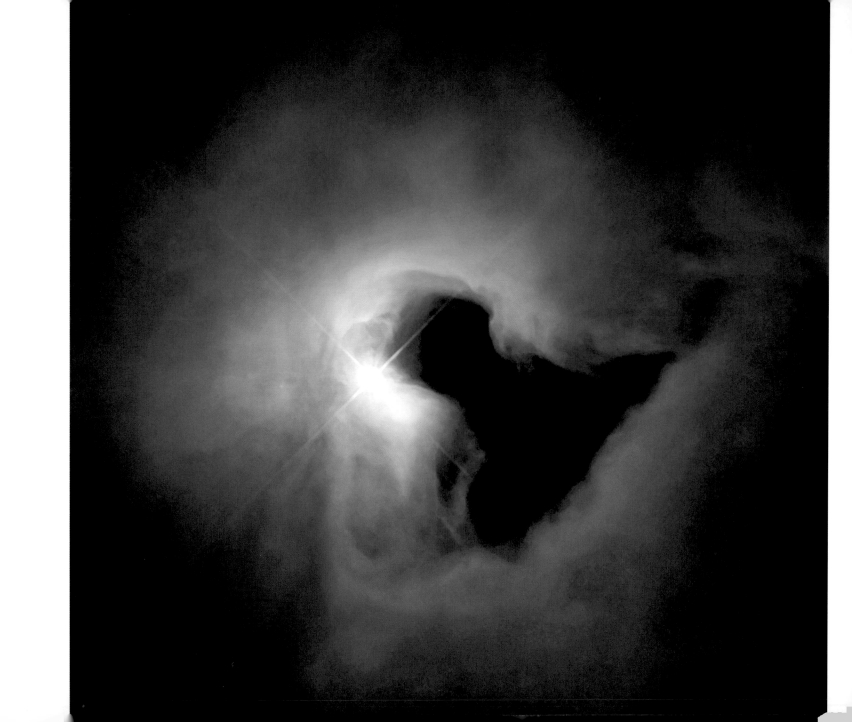

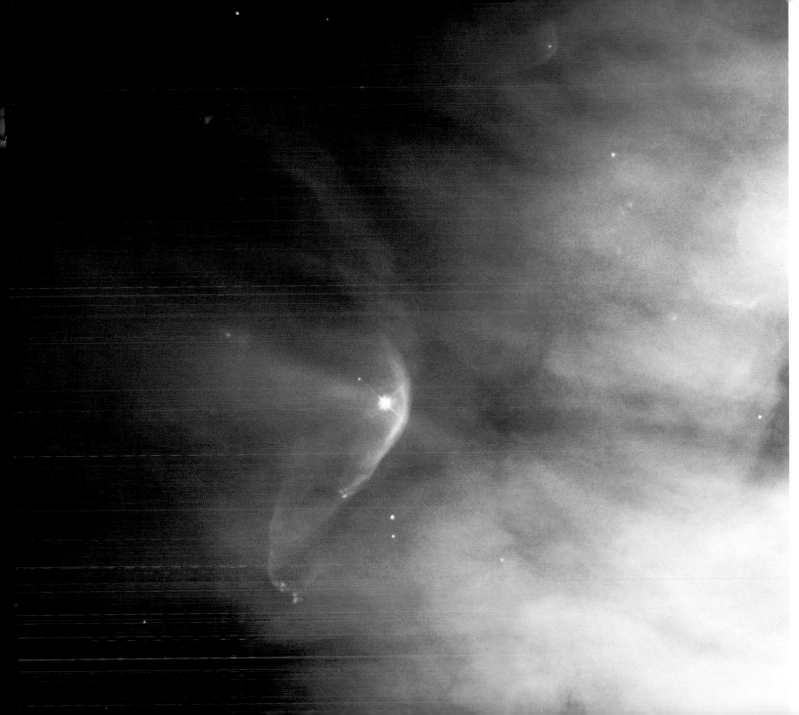

LEFT: The Hubble Space Telescope continues to reveal various stunning and intricate treasures that reside within the nearby, intense star-forming region known as the Great Nebula in Orion. One such jewel is the bow shock around the very young star, LL Ori, featured in this Hubble Heritage image.

FAR LEFT: Just weeks after NASA astronauts repaired the Hubble Space Telescope in December 1999, the Hubble Heritage Project snapped this picture of NGC1999, a nebula in the constellation Orion. The Heritage astronomers, in collaboration with scientists in Texas and Ireland, used Hubble's Wide Field Planetary Camera 2 (WFPC2) to obtain the color image.

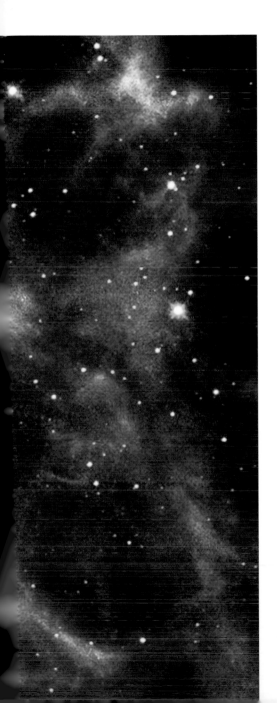

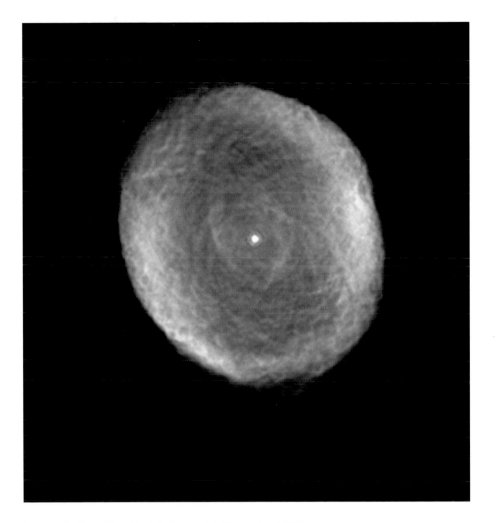

ABOVE: Glowing like a multifaceted jewel, the planetary nebula IC418 lies about 2,000 light-years from Earth in the constellation Lepus. In this picture, the Hubble telescope reveals some remarkable textures weaving through the nebula. Their origin, however, is still uncertain.

LEFT: In this unusual image, the HST captures a rare view of the celestial equivalent of a geode—a gas cavity carved by the stellar wind and intense ultraviolet radiation from a hot young star. Real geodes are baseball-sized, hollow rocks that start out as bubbles in volcanic or sedimentary rock. Only when these inconspicuous round rocks are split in half by a geologist do we get a chance to appreciate the inside of the rock cavity that is lined with crystals. In the case of Hubble's 35 light-year diameter "celestial geode," the transparency of its bubble-like cavity of interstellar gas and dust reveals the treasures of its interior.

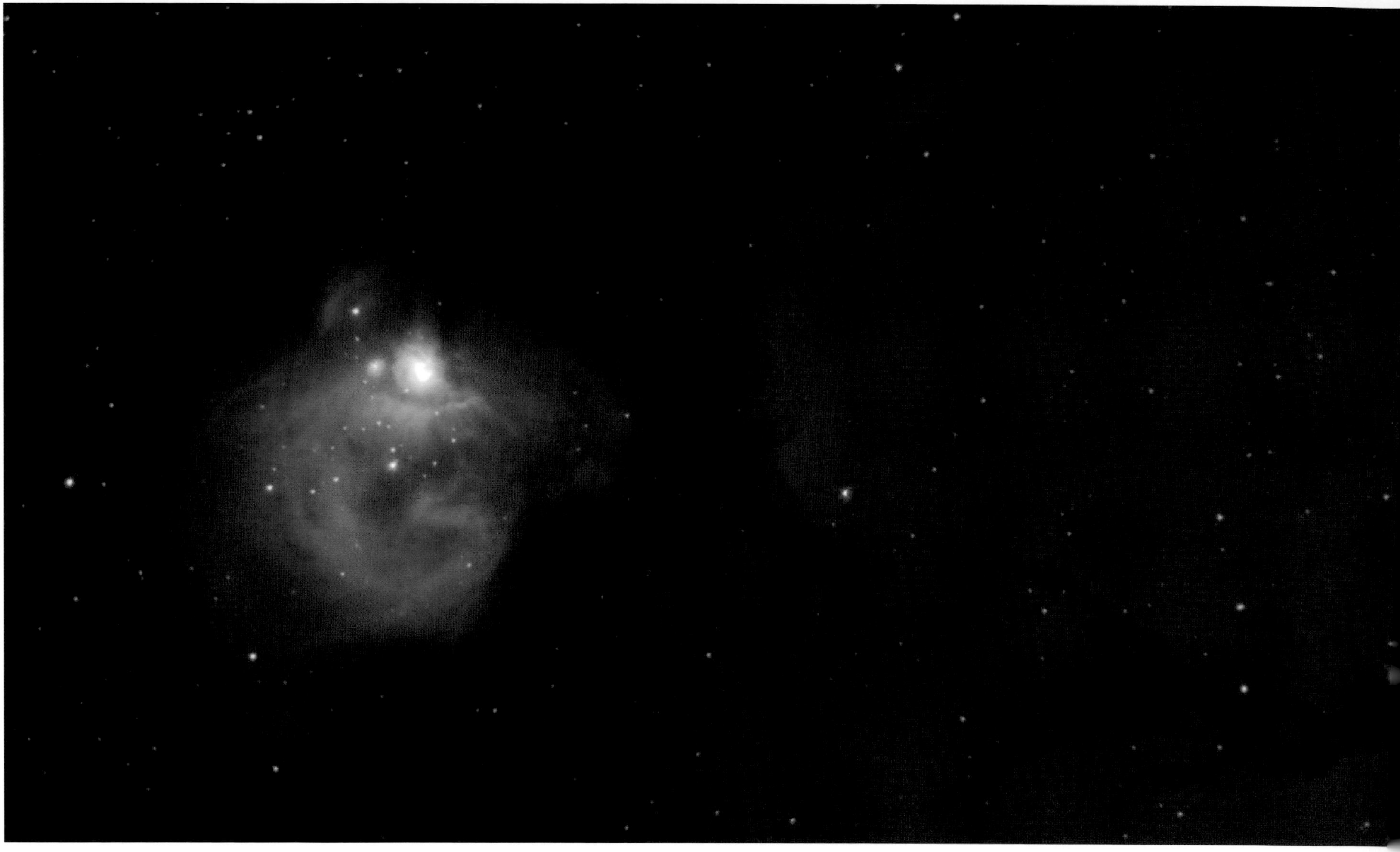

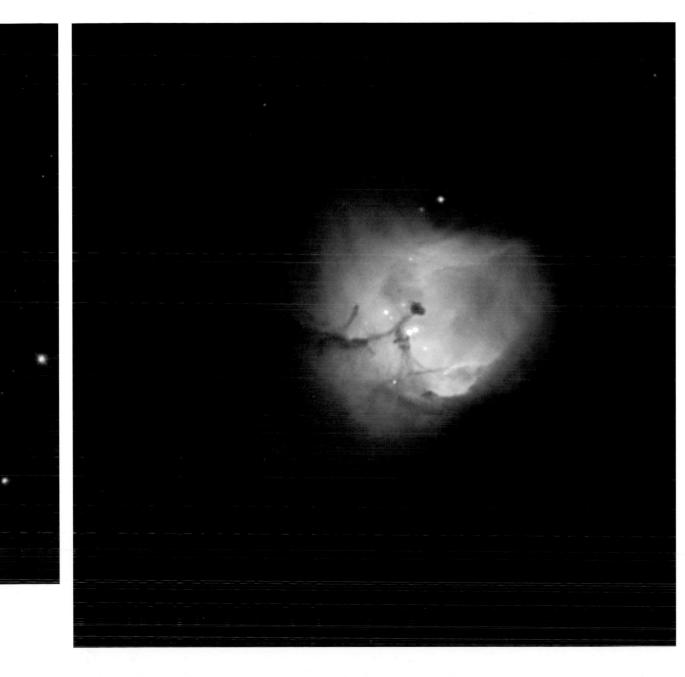

LEFT: The Hubble telescope has peered deep into a neighboring galaxy to reveal details of the formation of new stars. Hubble's target was a newborn star cluster within the Small Magellanic Cloud, a small satellite galaxy of our Milky Way. The picture shows young, brilliant stars cradled within a nebula, or glowing cloud of gas, cataloged as N01.

FAR LEFT: Extremely intense radiation from newly born, ultra-bright stars has blown a glowing spherical bubble in the nebula N83B, also known as NGC1748. A new Hubble telescope image has helped to decipher the complex interplay of gas and radiation of a star-forming region in the nearby galaxy, the Large Magellanic Cloud. The image graphically illustrates just how these massive stars sculpt their environment by generating powerful winds that alter the shape of the parent gaseous nebula. These processes are also seen in our Milky Way in regions like the Orion Nebula.

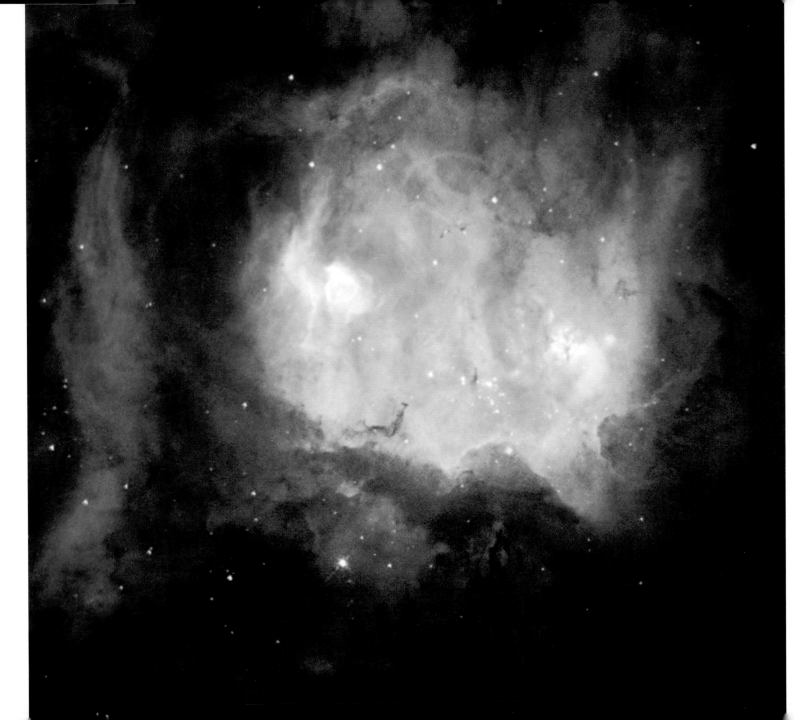

LEFT: The colors produced by the light emitted by oxygen and hydrogen help astronomers investigate the star-forming processes in nebulas such as NGC2080. Nicknamed the "Ghost Head Nebula," NGC2080 is one of a chain of star-forming regions lying south of the 30 Doradus Nebula in the Large Magellanic Cloud that have attracted special attention. These regions have been studied in detail with Hubble and have long been identified as unique star-forming sites. 30 Doradus is the largest star-forming complex in the whole local group of galaxies.

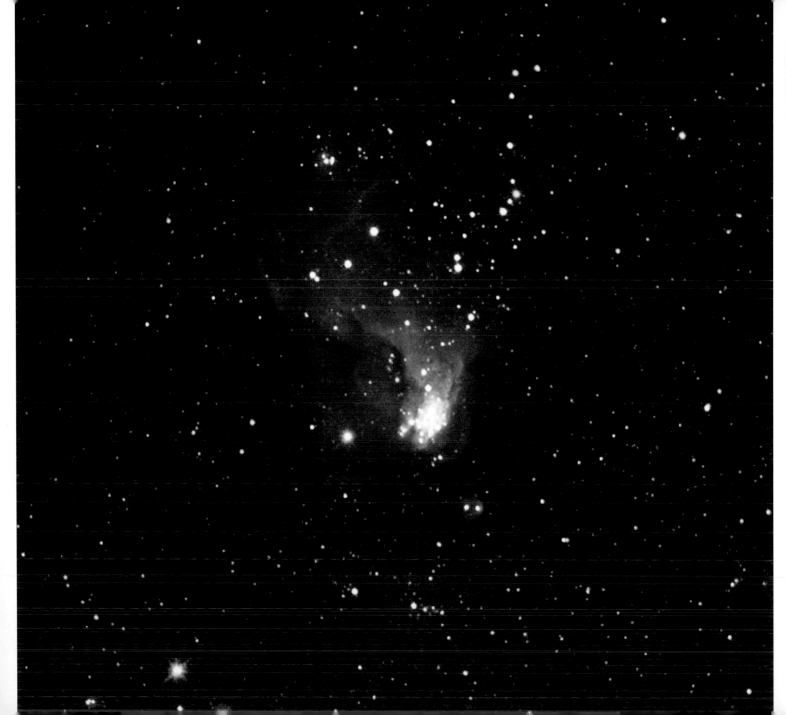

THE UNIVERSE | NEBULAS

LEFT: A revealing Hubble image showing Galaxy NGC7252, located 300 million light-years away in the constellation Aquarius. The spiral structure in the center is 10,000 light-years across while the whole area is 46,000 light-years across. The concentration of globular star clusters to form the central disc is unusual as they are normally spread more evenly throughout a galaxy, as in our own Milky Way. The star clusters are estimated to be between 50 and 500 million years old and are thought to have formed following a collision between two galaxies around one billion years ago.

Stars and Black Holes

The night sky is filled with stars, literally millions of them, and to the casual observer they all appear the same except for the fact that some appear brighter than others. Some of the brightest and most prominent have been given names, many of Arabic origin as the ancient civilizations in the Middle East were among the first to take a serious academic interest in the study of celestial objects, but Greek and Roman names are also common. Today stars are systematically cataloged and given reference numbers and letters—not so romantic, but practical when there are so many to be listed and studied.

The brightness of a star is referred to in terms of magnitude. Attempts to classify stars by this method date back to Greek times, although a more refined system, but still based on the ancient method, was established in the neineteenth century and is known as Pogson's scale after the astronomer who introduced it. Under this system a star of the 1st magnitude is two and a half times as bright as a 2nd magnitude star, which in turn is two and a half times brighter than a 3rd magnitude star, and so on. Stars less bright than 6th magnitude are difficult or even impossible to observe with the naked eye. On the other hand, objects brighter than 1st magnitude are rated as Zero (0) magnitude and even brighter ones are –1 or –2 and upward. On this scale the sun has a magnitude of –23.5 while the brightest star, Sirius, is –1.4. Unfortunately, this scale only relates to apparent brightness as we observe them from earth and distant star may actually be much brighter than a nearby one although it may not appear so to us. Consequently the concept of absolute magnitude was introduced under which the calculated brightness at a fixed distance of 32 light-years (10 parsecs) is used to determine magnitude. Under this method the sun rates as +5 while Rigel, the brightest star, rates as –6.8

Stars can also be classified in respect of their spectral properties (apparent colors) which provide an indication of their surface temperatures. This ranges from Type O, representing the brightest and hottest blue/white stars

down through yellow down to Type M (red). The complete set of letter designation is (from hottest to coolest) O, B, A, F, G, K, M. A plot of magnitude against spectral classification produces what is known as a Hertzsprung Russell diagram which is much utilised in analysing the characteristics of stars and star clusters in other galaxies.

It is easy to think of stars as permanent objects but in fact they do have a clearly defined life and examples of the various stages abound. For a start, they are not solid objects but very dense formations of gas, even the coolest having temperatures well in excess of 2,000°C (3,632°F)—more than enough to vaporize solid or liquid. Basically a star is born when gravitational forces are strong enough to condense a cloud of dust and gas to form a body of material which is massive enough for a process of nuclear fusion to begin at its core. This is caused by the tightly packed atoms colliding with each other and generating incredible amounts of heat. Such young stars burn very brightly but their rate of burning depends on their mass: high mass stars burn hotter and faster than smaller ones which fall into the lower spectral classifications. Over a period of time, all stars will gradually use up the hydrogen fuel and begin to burn out. At this stage they begin to throw off debris and contract to become a white dwarf, a small object of incredible mass. Even smaller and more dense objects formed from collapsing stars are known as neutron stars or pulsars. Whether a star forms a white dwarf or pulsar depends on its original mass, but if it is roughly equivalent to twice the mass of our sun then it will collapse into such a dense and small object that its gravitational effect will prevent any energy, including light, escaping from its influencer and literally forming a black hole. The dust and debris thrown out by a star in its dying stages drifts into space and forms massive nebula which then provide the raw material for the creation of further stars, often under the gravitational effects of white dwarfs, and black holes, thus furthering a continuing cycle of regeneration.

RIGHT: Wonderful view of the Pleiades, named after the seven sisters of antiquity, or Messier 45. In fact there are hundreds of stars in the star cluster but only a few are visible to the naked eye. They lie 425 light-years from Earth in the constellation Taurus and are thought to have formed 100,000,000 years ago.

LEFT: In October 1604 observers on Earth were startled to see a new bright star in the sky. This was just before the invention of the practical telescope and consequently modern observers are today able to view the remnant of what became known as Kepler's supernova in much greater detail. Using a combination of X-ray, infrared, and visible-light sensors, a composite image has been built up which shows a bubble-shaped shroud of gas and dust which is now 14 light-years wide and still expanding at a rate of four million miles per hour. The shockwave spreading outwardsheats and compresses gas clouds causing them to glow and define the shape of expanding remnants of the original explosion some 400 years ago.

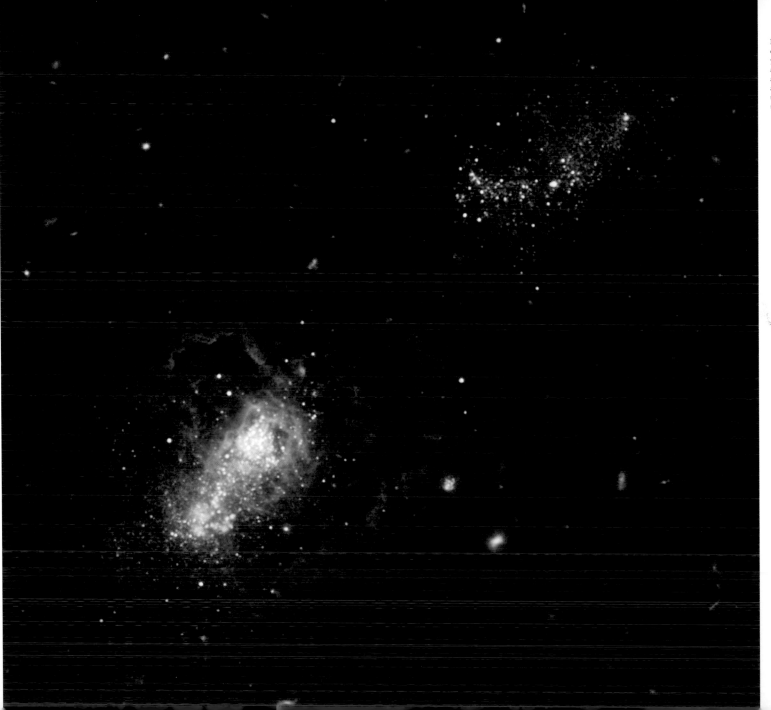

LEFT: This could be the youngest galaxy ever seen in the universe—at 500 million years old. Called I Zwicky 18, after the Swiss astronomer Fritz Zwicky, it did not start active star creation until 13 billion years after the Big Bang. In terms of the age of the universe, that's 20 times younger than the Milky Way.

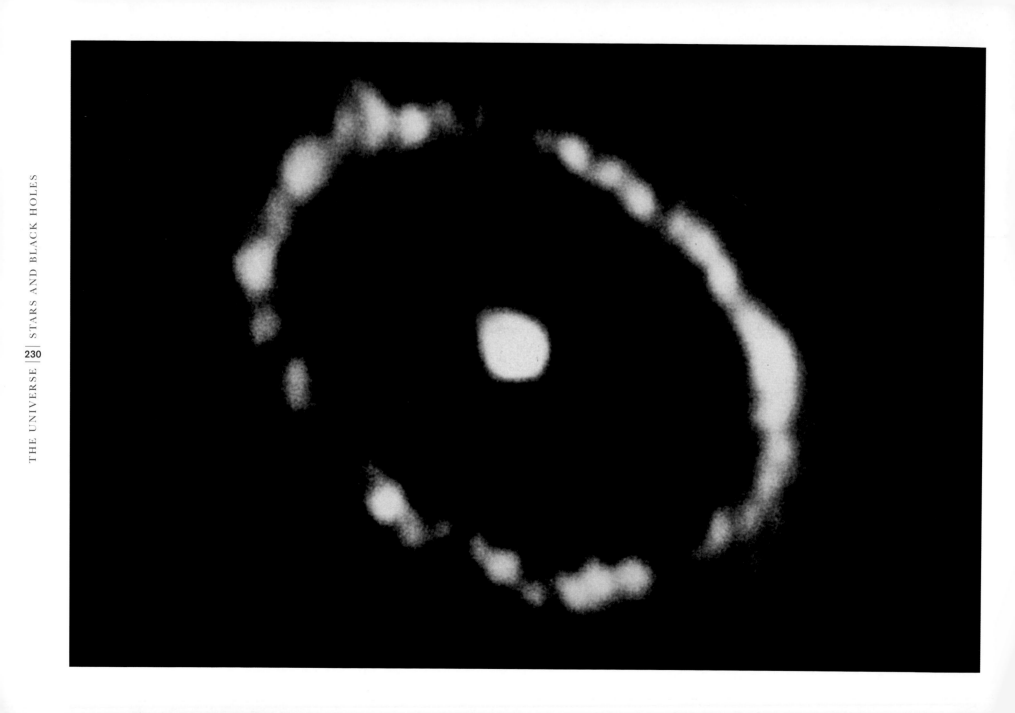

LEFT: A spectacular view of a recent supernova which exploded in the Large Magellanic Cloud (LMC) on February 23, 1987, taken by the European Space Agency's Faint Object Camera. Taking the designation Supernova 1987A after the year of its occurrence, its blast illuminated a ring of material which had been produced by an earlier explosion several thousand years ago. Not previously visible, the ring now glows as a result of being heated to 20,000°C (36,000°F) by the supernova explosion. The ring is inclined at 43 degrees to the line of sight and the light from the farthest edge reaches earth 1.37 years after the light from the forward edge. These and other measurements enable astronomers to accurately fix the distance from Earth of the LMC as 169,000 light-years.

RIGHT: A high-resolution view of the globular cluster 47 Tucanae which provides a fascinating insight in the possibility that stars may collide and capture each other, and in doing so recharge themselves and lengthen their lifespans. This image reveals a higher than expected concentration of a class of stars known as blue stragglers. Conventionally globular clusters were formed some 15 billion years ago and therefore only older stars would be expected. However, the revelation through Hubble images of a core of much brighter and, apparently, newer stars has led to the theory that the gravitational effects of stars coming in close proximity is enough to rekindle the nuclear fusion process and hence the rate of hydrogen burn-off.

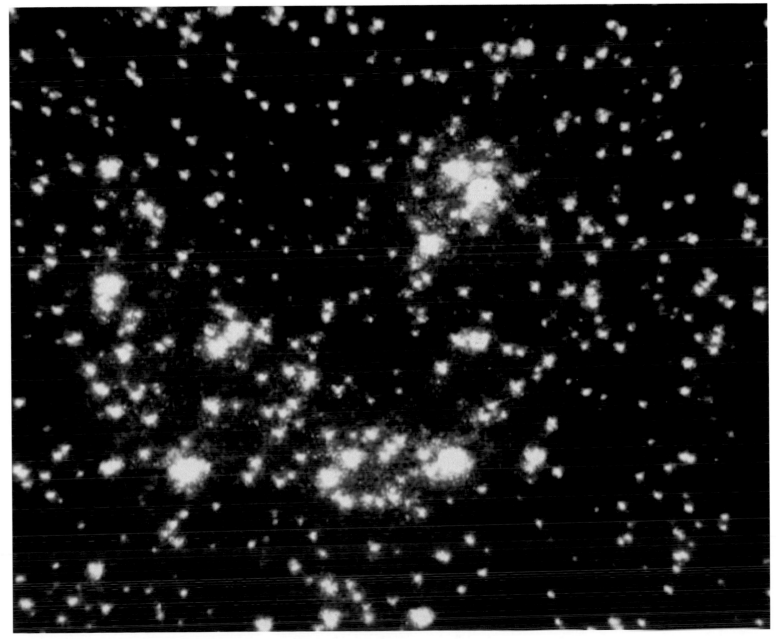

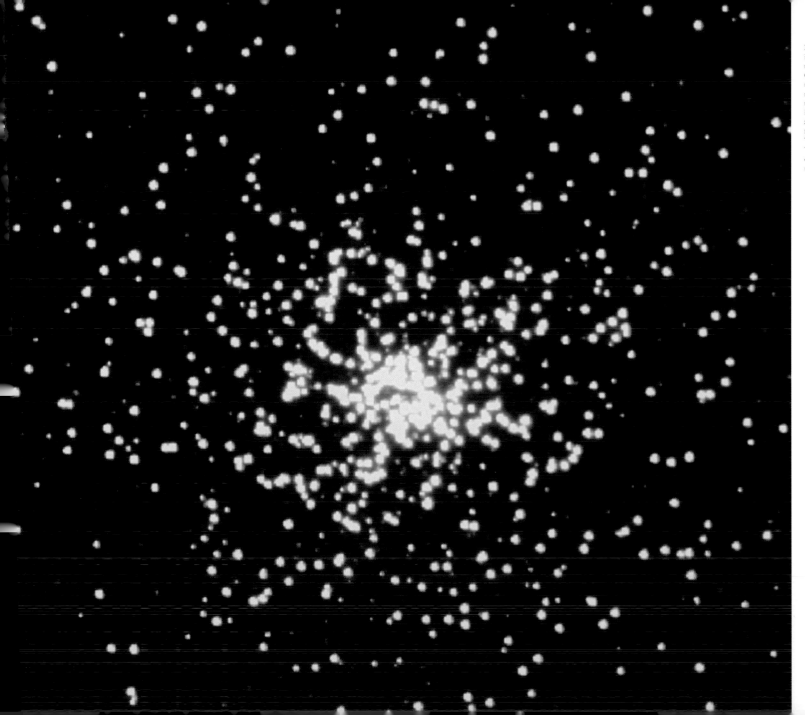

LEFT: A dramatic example of the improvement obtained in images from the Hubble's Wide Field and Planetary Camera 2 following its installation to replace the earlier WFPC1 during the 1993 repair mission. On the right is a WFPC1 image of the globular cluster M15 in which the individual stars at the center are merged into a single mass and even the outer ones suffer from the halo effect of spherical aberration. The other image was taken in April 1993 by WFPC2 and the individual stars can now be clearly differentiated while several previously unseen faint stars are now visible.

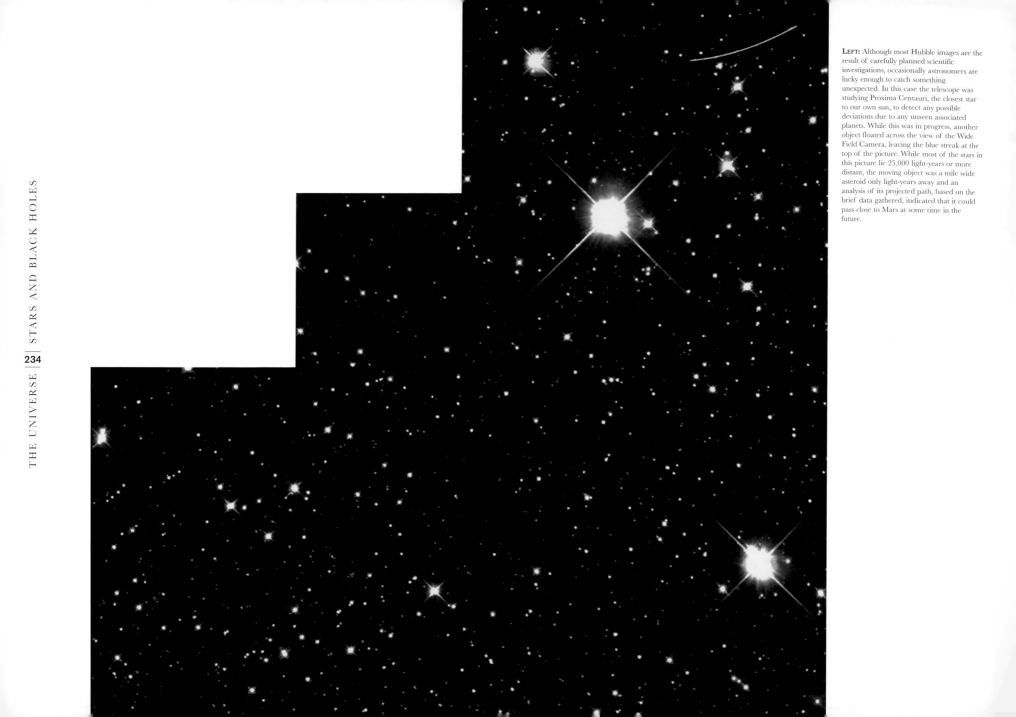

LEFT: Although most Hubble images are the result of carefully planned scientific investigations, occasionally astronomers are lucky enough to catch something unexpected. In this case the telescope was studying Proxima Centauri, the closest star to our own sun, to detect any possible deviations due to any unseen associated planets. While this was in progress, another object floated across the view of the Wide Field Camera, leaving the blue streak at the top of the picture. While most of the stars in this picture lie 25,000 light-years or more distant, the moving object was a mile wide asteroid only light-years away and an analysis of its projected path, based on the brief data gathered, indicated that it could pass close to Mars at some time in the future.

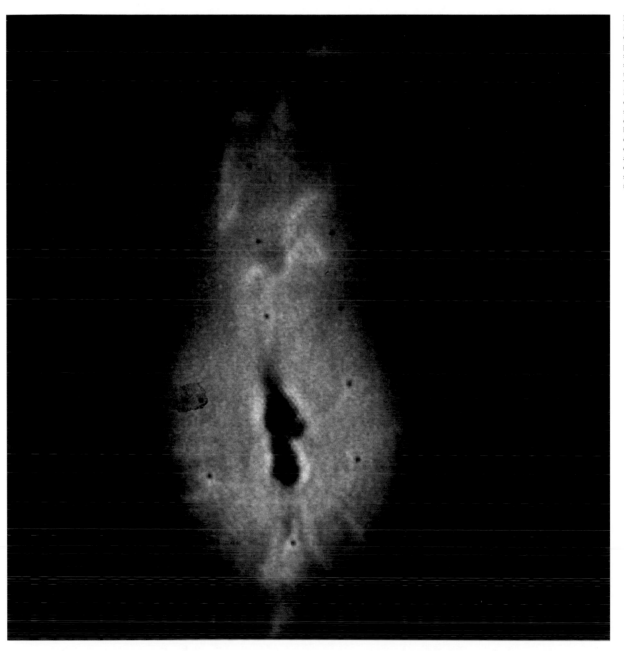

LEFT: Almost as soon as it was launched, the Hubble telescope was sending back new and detailed images to Earth despite the problems with the main reflector mirror. A striking example is this view of the symbiotic star system R-Aquarii taken during the telescope's first year of operation, 1990. The two dark knots at the center of the image contain the binary star system itself consisting of a red giant and white dwarf star (they appear in this way as the camera detector system was saturated by the very bright objects). It is thought that the white dwarf star is reenergized by large quantities of material drawn to it from its neighbor and the filament-like structures emanating outward in a spiral pattern are plasma streams of hot gas concentrated by strong magnetic fields.

RIGHT: A closeup view of the spectacular star cluster at the center of the nebula 30 Doradus. The large blue area to the left of center is the cluster R136 which contains several dozen of the most massive stars known up to 100 times the mass of our own sun and probably formed some two million years ago, making them very young in the time scale of the universe. The 30 Doradus Nebula has been an object of great interest due to the insight it has given astronomers into the origins of stars and the universe. This composite image using overlapping pictures taken between 1994 and 2000 and shows previously unseen pillars of dust and gas orientated toward the central cluster which are similar in size to those observed in the Eagle Nebula.

LEFT: A picture of the giant galactic NGC3603 Nebula which manages to capture several stages of star development in one view. The most prominent object is the blue supergiant Sher 25 at upper right center which has a unique circumstellar ring of glowing gas. Near the center of the picture is a starburst cluster which includes several hot Wolf-Rayet and early O-type stars. The intense ionizing radiation from this cluster has burnt a cavity in the center of the surrounding gas clouds and has formed giant gaseous pillars while dark clouds at upper right contain so called Bok globules. The complete progression runs through the globules and pillars to the evolving massive stars in the starburst cluster while the blue supergiant represents the ending of the life cycle.

RIGHT: Yet another image from that fruitful area of the southern sky, the Large Magellanic Cloud. In this case it is the double-star cluster NGC1850—the brightest such formation in the LMC after the well known 30 Doradus complex. The main cluster in this picture is estimated to be 50 million years old while the smaller cluster at lower right is only (!) four million years old. Of special interest in the smaller cluster are some faint red T-Tauri stars. These are very young solar-sized stars which are still forming and have not yet started converting hydrogen to helium, the process by which our own sun produces its radiant energy.

LEFT: The globular star cluster 47 Tucanae (see photo at right) located 15,000 light-years away in the southern constellation Tucana is being studied by astronomers for evidence of conditions suitable for planet formation and evolution.

These myriad points of light are part of the million or more stars which make up the globular cluster 47 Tucanae set in the constellation Tucana (the Toucan) over 15,000 light-years from Earth. Surveys by astronomers have revealed the fact there appear to be no planets to any of these stars, an unusual state of affairs which indicates that the system's environment is too hostile to their inception or that it lacks the elements to make them. There is a preponderance of older red and yellow stars with few young and hot blue stars leading astronomers to surmise that the cluster was formed over 10 billion years ago.

RIGHT: A fascinating image of the star Fomalhaut (at the center of the picture) which characterized by a dusty ring shaped belt which has a radius at its outer edge of around 12 billion miles. Examination of the Hubble image reveals that the inner edge of the ring is much sharper than its outer edge. The most likely cause of this effect is the presence of a planet whose gravitational effect is acting like a brush as it orbits the star. Fomalhaut is only 25 light-years away from the sun in the southern constellation Pisces Austrinus. The star itself has been shaded out in this image so that the relatively faint ring could be more clearly seen.

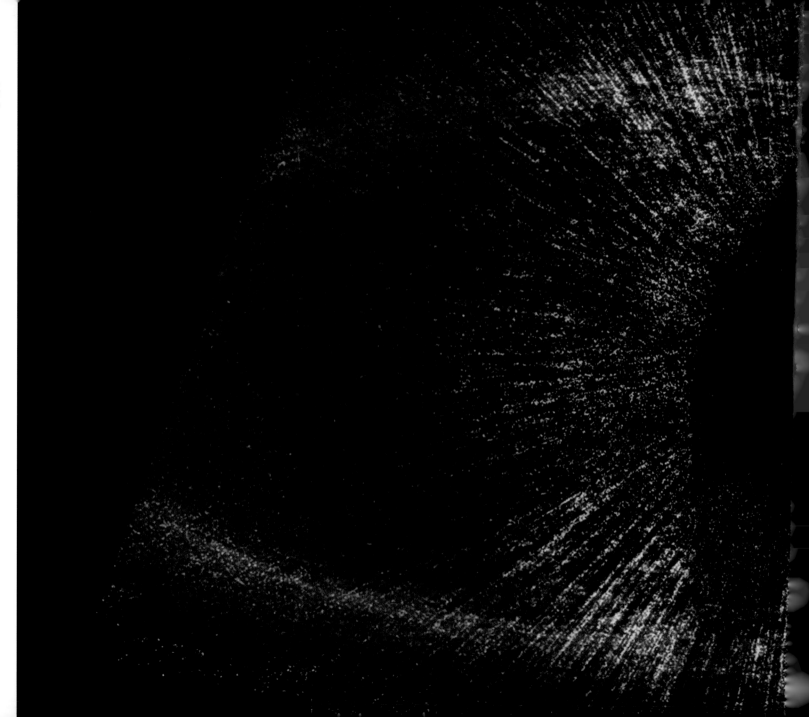

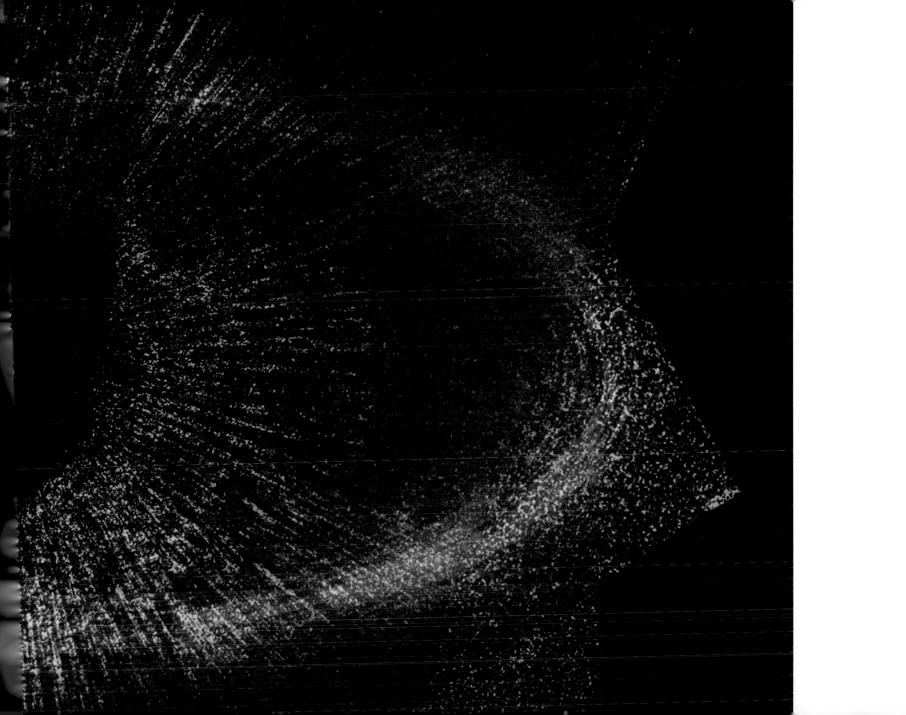

LEFT: Lying many millions of light-years away in the Virgo cluster of galaxies, this gigantic spiral is designated M100. The launch of the Hubble telescope, and particularly after its initial defects were repaired, allowed astronomers to study individual stars in this galaxy for the first time. In particular measurement of the brightness values of Cepheid variable stars is providing data-enabling accurate calculations to be made in respect of the distance of various other objects in the universe. Prior to Hubble, the Cepheids were to faint and the resolution too poor to allow them to be viewed individually.

LEFT: A massive supernova explosion was observed in the year 1572 by the Danish astronomer Tycho Brahe and other observers of the time. This image shows the debris from this event as it appears today and provides evidence to support the theory that certain types of supernova are result from a binary-star system consisting of a white dwarf and a normal star. The dwarf eventually explodes after being overfueled by the companion star which is then thrown away by the resulting explosion. A detailed examination of this image shows a star whose position and motion relative to the center of the burst makes it highly likely that this has in fact occurred, the example in question moving at three times the speed of other stars in its vicinity.

Glossary

A

ABSOLUTE ZERO

The coldest possible temperature when all molecular motion stops. This is zero on the Kelvin scale, –273°C or –460°F.

ABSOLUTE (APPARENT) MAGNITUDE

A scale that measures the inherent brightness of a celestial object by the apparent brightness or magnitude it would have if it were located exactly 32.6 light-years (10 parsecs) away. The lower the number (negative numbers), the brighter the object. (See also Bolometric Magnitude and Magnitude.)

ABSORPTION

The process by which light transfers its energy to matter. It leads to Absorption Lines—dark lines in a continuous spectrum caused by absorption of light. Each chemical element emits and absorbs radiated energy at specific wavelengths, making it possible to identify the elements present in the atmosphere of a star or other celestial body by analyzing which absorption lines are present.

ACCELERATING UNIVERSE

A model for the universe in which the attractive force of gravity is counteracted, driving matter apart at speeds that increase with time.

ACCRETION DISK

A flat rotating disk of gas and dust in space that surrounds a center of gravitational attraction—a black hole, a newborn star, or any massive object that attracts and swallows matter—that is growing by attracting matter to it. As the gas spirals in, it becomes hot and emits light or even X-radiation.

ACTIVE GALAXY

A galaxy that produces huge amounts of energy, including Seyfert galaxies, quasars, and blazars. Galaxies that have a massive black hole at their center are called active Galactic Nuclei (AGN). They produce energy at all wavelengths of the electromagnetic spectrum.

ADAPTIVE OPTICS

A process in which distortions, such as those caused by the Earth's atmosphere, are removed from a telescope's image.

ADVANCED CAMERA FOR SURVEYS (ACS)

An optical camera aboard the Hubble Space Telescope whose wavelength range spans from ultraviolet to near-infrared light. It was installed the camera aboard the telescope in March 2002.

ALBEDO

The ratio of the amount of solar radiation reflected from an object to the total amount it receives: so a high albedo (1.0) is shinier than a low albedo (0.0).

ANDROMEDA GALAXY

Also known as M31 and NGC224, Andromeda is our closest major galaxy. A spiral galaxy in our Local Group (see below), it is some 700,000 parsecs from the sun and the most distant object visible to the unaided eye.

APPARENT MAGNITUDE

See Absolute Magnitude. (See also Bolometric magnitude and Magnitude.)

ASTEROID

Small, rocky objects, most of which orbit the Sun. In the solar system most are in the asteroid belt between Mars and Jupiter. Ceres is the largest known asteroid with a diameter of 579 miles.

ASTRONOMICAL UNIT (AU)

Equal to the average distance from the Earth to the Sun, about 93 million miles or about 8.31 light minutes; a convenient unit for measuring distances between planets and their stars.

Planet	AU distance from the Sun
Mercury	0.39
Venus	0.723
Earth	1.0
Mars	1.524
Jupiter	5.203
Saturn	9.539
Uranus	19.18
Neptune	30.06
Pluto	39.53

ASTRONOMIC SCINTILLATION

The twinkling (fluctuation of intensity) of stars seen through a planet's atmosphere; also known as stellar scintillation.

ATMOSPHERE

The atmosphere is the mixture of gases that surrounds a planetary object, moon, or star.

AZIMUTH

The angle of the object from the observer's north point (projected onto the horizon). If an object is due north, its azimuth is 0 degrees. If it is due east, its azimuth is 90 degrees, etc. To find an object in the sky, two coordinates are needed, its altitude and its azimuth.

B

BARRED SPIRAL GALAXY (SB)

A spiral galaxy whose center is a bar-shaped band of stars and interstellar matter. The Milky Way is thought to be a barred spiral galaxy.

BAYER SYSTEM

This system—named after Bavarian Johannes Bayer (1564-1617)—assigns Greek letters (alpha for the brightest, then beta, gamma, delta, epsilon, zeta, eta, theta, iota, kappa, lambda, mu, nu, xi, omicron, pi, rho, sigma, tau, upsilon, phi, chi, psi, and omega) to stars in every constellation.

BIG BANG THEORY

The theory explains the start of time and the universe. It says that it began as an explosion of space-time 14 billion years ago. At that point everything that is in our universe was compressed in a point infinite density and has been expanding ever since.

BINARY STAR

Two stars that rotate around their common center of mass, bound together by their mutual gravitational attraction. About half of all stars are in a group of at least two stars; the fainter of the two stars is called the companion.

BLACK DWARF

A small, dense, cold, dead star—what remains after a red giant (see below) loses its outer layers, forming a planetary nebula and then a white dwarf. Our sun will turn into one.

BLACK HOLE

Coined by the physicist John Archibald Wheeler, a black hole is a massive object (or region) in space that has collapsed under its own gravitation and is so dense that within a certain radius (the event horizon, see below) its gravitational field does not let anything escape from it, not even light. It is thought that giant stars will evolve into red supergiants, then supernova , and then black holes. Some astronomers think that there may be a black hole at the center of each galaxy.

BLAZAR

A type of quasar—a distant star-sized energy source—blazars emit jets of gamma rays and other electromagnetic radiation.

BLUE GIANT STAR

A rare, huge, hot, luminous, blue star such as Rigel and Regulus.

BOLOMETRIC MAGNITUDE

This measurement includes a star's entire spectrum of radiation, not just the visible light. (See also Absolute Magnitude and Magnitude.)

CARTWHEEL GALAXY

The Cartwheel Galaxy—about 500 million light-years from Earth in the constellation Sculptor—was a spiral galaxy but it has been hit by a smaller galaxy. The result has a ring-like structure.

CEPHEID VARIABLE STARS

Supergiant stars—such as Polaris or Delta Cephei— whose luminosity varies in a periodic fashion. As they get bigger, they lose brightness; then they get smaller and increase in brightness. The period over which the star varies is directly related to its light output, so its absolute magnitude (see above) can be calculated from the observed period, which can then give the distance to the star.

CLUSTER

A group of stars that can number in their thousands. Galactic clusters, sometimes called open clusters, contain up to a few hundred; globular clusters (see below) contain tens of thousands.

COLLIDING GALAXIES

As its name suggests, this is when two galaxies pass close enough to disrupt each other, and can result in the galaxies merging. (See also Cartwheel Galaxy.)

COMA

The gas that surrounds a comet (see below).

COMET

A body that orbits the Sun. It is made up of a solid nucleus, a gaseous coma, and a tail of dust and gases that points away from the sun, because of the force of the solar wind.

CONSTELLATION

The groupings of stars as they are seen from Earth. The earliest (northern hemisphere) constellations were the 48 named in antiquity after such things as mythological figures (Andromeda, Cassiopeia, Orion, Perseus) or animals (Aquila—the eagle, Leo—the Lion, Ursa—the bear). More were added from the southern hemisphere to give a total of 88 constellations. In modern usage, each constellation incorporates a precisely defined region of the sky.

CONSTELLATION FAMILY

Constellations that are either close to one another or have some other relationship—such as the 12 Zodiac constellations or the 10 in the Ursa Major family.

CORONA

The hot outer layer of a star's atmosphere visible during a solar eclipse that can extend millions of miles from the surface.

COSMIC BACKGROUND RADIATION

Created at the time of the Big Bang, it has since cooled to a temperature of 3 K.

DEFERENT

A Ptolemaic term, the deferent is the large circular orbit (it would be the seventeenth century before Kepler discovered that orbits were elliptical) around which a planet was thought to orbit in one or many epicycles (see diagram on page 18).

DIFFUSE NEBULA

A wide, spread-out, irregularly-shaped cloud of gas that can be up to 100 light-years wide.

DOPPLER SHIFT

Named for Johann Christian Doppler (1803–53), this is the increase or decrease in wavelength of an object that emits a wave (light, sound, etc.) as the object moves relative to the observer. Things moving toward you have their wavelengths shortened. Things moving away have their emitted wavelengths lengthened, such as a star traveling away from Earth, whose light appears redder (the light waves are elongated, lengthening the wavelength)—the red shift (see below). The expansion of the universe was discovered when Hubble observed that the light from almost all other galaxies was red-shifted.

EAGLE NEBULA

The Eagle Nebula (M16) is about 7,000 light years from Earth in the constellation Serpens. This star-forming cloud is illuminated by ultraviolet light that is emitted from newborn stars.

ELECTROMAGNETIC RADIATION/ SPECTRUM

Energy in wave form shown below in order of decreasing energy and increasing wavelength.

NEUTRON STAR

An extremely compact ball of neutrons created from the central core of a star that collapsed under gravity during a supernova explosion. They are extremely dense compressing the mass of an average star into a ball only 10 kilometers or so in size. Neutron stars that regularly emit pulses of radiation are known as pulsars.

NOVA

A nova is produced in a binary system where hydrogen-rich matter is transferred onto a white dwarf. The star rapidly brightens, then slowly fades back to normal.

O

OLBERS PARADOX

Olbers Paradox (see page 48) is "Why is the sky dark at night when there are so many stars?"

ORION NEBULA

The Orion Nebula (M42 and M43) is a huge, nearby, turbulent gas cloud lit up by bright, young hot stars developing within the nebula.

P

PARALLAX

The apparent change in the position of a star that is caused by the motion of the Earth as it orbits the Sun (see pages 52–53). Friedrich Bessel first detected this in 1838 observing the star 61 Cygni.

PARSEC

A unit of distance that is equal to 3.262 light-years or 3.085678×10^{13}

kilometers. It is the distance at which a star would have a parallax of one second of arc.

PLANETARY NEBULA

An expanding shell of glowing gas expelled by a star late in its life. Our Sun will create a planetary nebula at the end of its life.

PULSAR

A pulsar is a rapidly spinning magnetized neutron star that emits rapid and periodic pulses of energy— radio, visible, x-ray and gamma radiation—in pulses. Discovered in 1967 by S. Jocelyn Bell Burnell (1943–), who controversially did not share in the 1974 Nobel Prize for the discovery. The pulse is produced every time the rotation brings the magnetic pole region of the neutron star into view.

Q

QUASAR

A quasar—quasi-stellar radio object because this type of object was first identified as a type of radio source —is a distant star-sized energy source (i.e. unresolved object) in space with excess of ultraviolet located at very large distances from us (as indicated by their high redshifts). First detected by Maartin Schmidt and Allan R. Sandage in 1963–64. Quasars are believed to be powered by supermassive black holes in the centers of galaxies.

R

RADIO ASTRONOMY

Exploring space by examining radio waves. Pioneered by Karl Gothe Jansky: in 1932 he was the first to detect radio waves from a cosmic source.

RADIO GALAXY

Any galaxy whose luminosity is greatest in radio wavelengths.

RED DWARF

A small, cool, very faint, main sequence star—the most common type of star in the universe—such as Proxima Centauri.

RED GIANT STAR

A relatively old, post-main-sequence star whose diameter has swollen and whose temperature has cooled (thus red) such as Betelgeuse. Our Sun will become a red giant in about five billion years.

RED SHIFT

The red shift is lengthening of a light wave that is emitted from an object that is moving away from us (see also Doppler Shift)—for example, when a galaxy is traveling away from Earth. This increase in wavelength makes the object appear to be redder than it actually is.

REFLECTION NEBULA

A nebula— frequently bluish in color—that glows as the dust in it reflects the light of nearby stars. A reflection nebula surrounds the Pleiades cluster.

RELATIVITY

Formulated by Einstein (see pages 54–55), the Theories of Special and General Relativity changed our understanding of space and time. Special Relativity is based on the idea that nothing can pass the speed of light but that time and distance measurements are relative. General Relativity expands the theory of special relativity to include acceleration and gravity.

S

SIDEREAL TIME

Time measured relative to the stars instead of relative to the motion of the Sun. One sidereal day is equal to 23 hours and 56 minutes and is the period during which the earth completes one rotation on its axis. A sidereal month is 27.322 days long. A sidereal year 365 days, 6 hours, 9 minutes, and 9.5 seconds long.

SMALL MAGELLANIC CLOUD
See Magellanic Clouds above.

SOLAR PLUME
A long, feathery jet of high-speed electrified gas that is expelled from the Sun's corona.

SOLAR SYSTEM
A solar system is a group of asteroids, comets, planets and moons held together (orbit around) by the sun's gravitational influence. In our solar system, nine planets, over 61 moons, and many other objects orbit our Sun.

SPACE TELESCOPE IMAGING SPECTROGRAPH (STIS)
Astronauts installed STIS aboard the Hubble Space Telescope in February 1997. STIS spans ultraviolet, visible, and near-infrared wavelengths. The spectrograph can sample 500 points along a celestial object simultaneously.

SPECTROSCOPY
A scientific technique in which the visible light coming from objects is examined to determine the their composition, temperature, density, and velocity.

SPEED OF LIGHT
According to Einstein, nothing can go faster than the speed of light which is defined as the speed electromagnetic waves can move in a vacuum: 186,000 miles/sec or 299,792,458 meters/sec.

SPIRAL GALAXY
As the names suggests, these are galaxies with a central bulge—typically of older stars—and surrounding spiraling arms of young, hot stars, as well as interstellar matter. There are two types—spiral and barred spiral. The Milky Way and Andromeda Galaxy are examples of spiral galaxies.

STAR CLASSIFICATION
Stars are classified by their temperature and their spectra (the elements they absorb). There are seven main types of stars. In order of decreasing temperature:

Type	Absorption
O	He II
B	He
A	H
F	Ca
G	strong metallic lines
K	bands developing
M	very red

For example, the Sun is a G1 star.

SUPERNOVA
The explosion of a star. This is caused either by a huge explosion at the end of a substantial star's nuclear-burning life or by sudden nuclear burning in a white dwarf star. A supernova releases a tremendous amount of energy (particularly light), expelling the outer layers of the star and becoming extremely bright. The star is destroyed and what remains is a white dwarf, a neutron star, or a black hole.

U

ULTRAVIOLET (UV) RAYS
Electromagnetic radiation with shorter wavelengths and higher energies and frequencies than visible light (see Electromagnetic Radiation/Spectrum). The ozone layer traps much of the Sun's ultraviolet energy coming through Earth's atmosphere. A UV telescope receives UV rays from space. Since the ozone layer traps much of the Sun's UV energy, orbiting UV telescopes are now used.

UNIVERSE
The totality of space and time, along with all the matter and energy in it. (See also Big Bang.)

V

VARIABLE STAR
A star whose luminosity (brightness) changes with time—the periods ranging from minutes to years.

W

WHITE DWARF
A small, very dense, hot remnant of a star near the end of its life. The final stage in the evolution of a star about the size of our Sun, it is what remains after a red giant loses its outer layers—after it has exhausted its sources of fuel for thermonuclear fusion. A white dwarf radiates mainly in the ultraviolet. Our Sun will turn into a white dwarf. The companion of Sirius is a white dwarf.

WIDE FIELD/PLANETARY CAMERA (WF/PC)
Originally eight separate, yet interconnected, cameras used as the main optical instrument on the HST. The Wide Field and Planetary Camera 2 replaced the WF/PC in December 1993. It takes photos of faraway objects and its 48 filters cover a wavelengths from ultraviolet to near-infrared light.

X

X-RAYS
A type of electromagnetic radiation between ultraviolet light and gamma rays in wavelength, frequency, and energy (see Electromagnetic Radiation/Spectrum).

PAGE 256: N63 is a star-forming region in the Large Magellanic Cloud. Visible from the southern hemisphere, the LMC is an irregular galaxy lying 160,000 light-years from our own Milky Way Galaxy. The LMC provides excellent examples of active star formation and supernova remnants to be studied with Hubble. See also photo on page 248.